@CivicWhitaker

Anthology

@CivicWhitaker Anthology

Three years of organizing, writing, and documenting in Chicago civic tech at the Smart Chicago Collaborative

CHRISTOPHER WHITAKER

Edited by Daniel X. O'Neil

To those in the trenches. And to my wife Tara for jumping in with me on this crazy ride.

Make no little plans; they have no magic to stir men's blood
and probably themselves will not be realized.
—Daniel Burnham

The benefits of education and of useful knowledge,
generally diffused through a community, are essential to the
preservation of a free government.
—General Sam Houston

Now, a lot of people have given up on government. And if you're
one of those people, I would ask that you reconsider, because things
are changing. Politics is not changing; government is changing.
— Jennifer Pahlka

The @CivicWhitaker Anthology: Three years of organizing, writing, and documenting in Chicago civic tech

by Christopher Whitaker
and the Smart Chicago Collaborative

Manufactured in the United States of America by the
Smart Chicago Collaborative
http://www.smartchicagocollaborative.org/ / @smartchicago
The Chicago Community Trust
225 North Michigan Avenue
Suite 2220
Chicago, IL 60601
(773) 960-6045

Made possible by funds from The John S. and James L. Knight Foundation, John D. and Catherine T. MacArthur Foundation, and The Chicago Community Trust

Set in Scala and ScalaSans

Library of Congress Control Number [applied for on Monday, September 21, 2015]
ISBN: 978-0-9907752-5-6

First Printing, September 2015

Contents

Introduction

Hiring Christopher Whitaker to work as a consultant for Smart Chicago was one of the best decisions I made here.

Together, we created a new job type— part documenter, part organizer, part evangelist, part original writer and thinker about an emerging subsector of the technology industry— civic tech.

Through our work together, he's helped build one of the strongest civic hacking communities in the country, been an essential part of the growth of the largest network of civic tech volunteers in the world, helped make the first weekend in June a national day of civic hacking, worked with a dozen emerging companies and organizations to grow revenue and impact, and served as a critical thread in the national fabric of this important movement.

This book is a simple anthology of the best of his vast work.

Daniel X. O'Neil
September 2015
Chicago

Part One

The Chicago Civic Innovation Ecosystem

Here's an introduction to Chicago's civic innovation ecosystem and some history of civic innovation events in Chicago.

Chicago's Civic Technology Ecosystem

This is the text of a speech I gave at the Civic Tech Forum in Tokyo, Japan describing Chicago's civic technology ecosystem. Much of the work that's been done in Chicago is possible because the ecosystem has set up the space for success. I talk about the different parts and how they work together.

Good afternoon, my name is Christopher Whitaker and I am a civic technology consultant from Chicago. It is an honor to speak with you today.

So first, a little about my hometown: Chicago is the third largest city in the United States and we're known for a few things. The first is that we're one of the coldest cities in the country – last winter we had 658 mm of snow. We're also known for the Chicago Cubs with their 100 year losing streak. We also are the home to deep dish pizza.... which comes to about 50mm of cheese, sauce and sausage. Our skyline sits on the shores of Lake Michigan, which showcases some of the best architecture in the county.

We're also known for being a civic innovation powerhouse. Our city has one of the biggest repositories of open data in the nation, we have one of the biggest weekly gatherings of civic hackers world-wide, and we are one of the most productive cities when it comes to projects that leverage the power of technology in the service of the public good.

We can do this because we've built an ecosystem design to power civic innovation. I'd like to tell you a little about my story and the story of our civic innovation ecosystem.

When I give my talks, I always like to ask how many developers are in the room. People who code, know Ruby, have GitHub Accounts. – Ok, so this is the part where I give out my deepest darkest secret that I am not a coder. I don't know Python, I don't know R and if you let me into your production branch on Github – I will destroy everything.

I came into civic technology from the opposite direction from a lot of my peers.

It was early 2009 and the American economy was in the worst recession in decades. Wall Street was crashing, houses were being foreclosed, and unemployment was skyrocketing. As part of the government's response, the federal government gave money to state governments to hire more bureaucrats to help process unemployment claims. I was hired by the State of Illinois in 2009 to work in a field office.

So, my first day of work I arrive 30 minutes early. It's early March and so it's still pretty cold in Chicago – and there is still a line of about 20 people out the door. I get to my desk and boot up my computer – and I wait about five minutes because the computer is from 1995 – It's got a big box monitor bigger than my head. I finally pull up the system that I would be using to help my friends and neighbors with their unemployment insurance...and I am greeted by a DOS screen.

It's got a green text, it sorta blinks at me. And I'm floored. Turns out this system was from 1975. The system wasn't a real time system. So, when I did anything (enter in somebody's claim for unemployment insurance, made sure the system sent them a check) – it didn't actually happen until it ran through an overnight process. The only way for me to know for certain if something I did worked was to come in the next morning and print a ream of paper about this thick (hand motion) – then go through line...by line...by line... to make sure that nothing rejected. Because if it rejected and nobody fixed it, the claimant would come in 3 days later asking us "Hey! You told me you fixed it! There's no money in my account – now I can't pay rent!"

At the same time, I'm carrying around an iPhone in my pocket.

Something 10 times more powerful that the system that I had to use to help my neighbors with their unemployment and I'm using the iPhone to watch cat videos on my way home. It was incredibly frustrating because to me it should be the opposite. As a front line public servant, I should have the best tools possible to help residents. So, I started getting more into technology as a way to try to find ways to do my job better. I started going to technology events, to hackathons, and looking for ways to get involved.

After graduating with a Master's of Public Administration from DePaul University, I became a consultant for the Smart Chicago Collaborative. Around the same time, I became one of Code for America's first Brigade Captains – and later became the Brigade coordinator for the Midwest region. I also help to co-host the Chi Hack Night in Chicago – the largest weekly gathering of civic technologists in the United States. I'm extremely privileged to be able be involved in so many of these efforts and to be part of this community. Specifically, I'm proud to represent Chicago's civic innovation community.

In the past few years, we've been able to accomplish quite a bit. Some examples of our recent wins include a site called Largelots.org. Because of the recession, Chicago has had a lot of abandoned buildings. The City tears the buildings down then takes ownership of the lot – which the city doesn't like to do because it's just sitting there. Teamwork Englewood, a community organization on Chicago's south side, worked with the city and a local civic technology company called DataMade to build a site that let's people buy these vacant lots for one dollar.

For another project, Smart Chicago worked with a local youth organization called Mikva Challenge to build an app (Expunge.io) to make it easier for people to get their juvenile records erased so that it's easier for people to find jobs and apply to university.

Smart Chicago also worked with our city health department to build an app that listens to Twitter for reports of food poisoning. So, if I eat somewhere and tweet that I'm sick, the Health Department will tweet me back asking for more information. That becomes a city service request and the city will send a restaurant inspector to that restaurant. We've done quite a bit of work – and we expect that work to continue as the community aims to take on bigger and more complex projects.

A big part of why we're able to have as much impact as we do is because we have built a civic innovation ecosystem. This ecosystem acts as an engine that helps power civic innovation and produce not only civic applications – but a more transparent government, non-profit community organizations are a more technologically savvy, more civic technology companies, and much more connected network of people who understand the dual challenges of both technology and government.

Each of these different parts complements the other. The City of Chicago provides the data that powers the engine. The Chi Hack Night provides the place for talent and collaboration. The Smart Chicago Collaborative provides institutional support.

And then, all of Chicago's efforts then get shared with the Code for America Brigade Network – which helps us collaborate on a global scale. I'd like to explain a little bit about each part in more detail.

First up, The City of Chicago likes to focus on the data. They view data as fuel: fuel for transparency, fuel for innovation, and fuel for businesses.

To help facilitate this, the city has undergone the process of linking all of their business system's directly to the city's data portal. So, when the city fills a pothole – the information is automatically uploaded to the data portal where it can be used by civic technologists. On the city employee side, they aren't doing anything different – there's no additional step they have to take. The OpenETL toolkit does the work automatically. The city has open sourced the OpenETL toolkit on GitHub – so other municipalities can do the same thing.

But, the most important thing the city does to help the ecosystem is their participation in the community. This is Chicago's Chief Data Officer Tom Schenk Jr at one of our Chi Hack Nights. Schenk is a regular attendee at the hack night and so when the community has questions about city data or needs a particular data set, we can simply ask him directly. The Chief Data Officer is also on Twitter at @ChicagoCDO – so anyone can reach out to him and ask questions. This community involvement is a necessity – it allows the city to respond to the community's data needs.

The next part of the civic innovation ecosystem that I want to talk about it people. None of these efforts that I've talked about work without people to come up with ideas, to build them, and to help implement them. And I'm not just talking about web developers – I'm also talking about city employees, designers, non-profit employees, data scientists, journalists – they're all needed to make our civic apps the best they can be.

The way we recruit people into our civic innovation space is through the Chi Hack Night. The Chi Hack Night was spun out of the OpenGov Chicago monthly meetup. The OpenGov Chicago Meetup was started in 2009 as a way to promote transparency in government. Once the City of Chicago began opening up their data, the challenge became "OK... so what do we do with all of this data?" The OpenGov Chicago meetups were a great place for talk – but not a great place to get things done. The city had started to host a series of hackathons, but we quickly discovered that hackathons were great for meeting other civic hackers but we really couldn't produce anything of quality at the hackathons.

So, Derek Eder started the Chi Hack Night – a weekly event for everyone to get together and work on civic apps and I started co-hosting a few months later. We've been doing the event for three years and it's become the place to talk about projects, challenges, and to meet like-minded people. We average about 80 people a week and have had up to 120 people at an event. If you want to learn about the civic technology space in Chicago – this is where you go. The most important part about building this community is making sure that everyone feels welcome and invited – not just the technologists – but community activists as well. Part of what I do at hack night is teaching the Civic Hacking 101 class so that it helps lower the barrier to entry. Nobody should feel intimidated when they come to join our community.

And it's very important to get everyone – all types of people – into the community. All of these wins that I spoke about earlier – only exist because we had a partnership form between the technologists and the people who are working in front-line organizations to help try to solve civic problems. The largelots.org project was spearheaded not by the coder – but rather the community organization that wanted to solve the problem of vacant lots inside their neighborhood. Expunge.io came about from a partnership between the legal aid foundation and a non-profit youth organization. There's an entire community organizing aspect to our work that provides tremendous value to what we do.

But that's a hard thing to do on a volunteer basis. There are only so many hours in a day to do this work if you're also working a full-time job. With a team of volunteers, it's easy to get small wins. For example, creating a searchable map of all the places where you can get a flu shot online. However, to do something more complicated and more complex – it requires people working full time. Being able to do that requires considerable funding.

In 2007, the MacArthur Foundation, the Chicago Community Trust, and the City of Chicago put together a report called the "City that Networks" The leadership of these organizations recognized that the digital revolution was poised to change the way that cities functioned. The leadership came together to find a way to set Chicago up for success. To do this, they decided that Chicago needed a center of gravity of civic innovation. That organization became the Smart Chicago Collaborative. The three organizations sit on its leadership board with the organization itself being houses as an initiative of the Chicago Community Trust. We receive funding from the MacArthur Foundation, and the City of Chicago sits on our board.

Because we have funding, the Smart Chicago Collaborative is able to hire consultants to build civic apps, meet with partners, and provide advice to other community organizations.

We're also able to add capacity to the ecosystem. Anyone who builds a civic app in Chicago can have it hosted by the Smart Chicago Collaborative for free. We also provide free user testing and access to developer tools like Twilio and Mapbox.

We've even been able to add capacity to our county government. Cook County wanted to hire a staff member to help open up their data. However, because of budget restraints they could not afford to hire somebody of adequate talent. So, Smart Chicago and Cook County split the difference enabling Kalov Strategies to start opening up Cook County data.

We have also been able to team up with World Business Chicago and neighborhood community organizations to help close the digital divide in Chicago through our Connect Chicago Challenge program.

This is only possible because we have the institutional support to help fund our efforts.

The other major activity that we do is writing about our work and to tell our story. Every week, I live stream the Chi Hack Night and blog about the presentations. I also write how-to guides and lessons learned to help newer civic innovators in the space. In the three years that I've been doing this for Smart Chicago, we've put together over 400 blog posts about civic innovation and have over 300 videos on YouTube. We've also written books about our work.

We do this because when other cities ask us how we did something, we want to be able to tell the story and to be helpful. As much work as we've done in this space, civic technology is still a fairly new field. None of this existed ten years ago. It's something that none of us went to school for. We've had to learn all of this through experience and doing the work. So, it's very important for us to write down what worked, what didn't, and what we were thinking at the time so that when others try and start their own projects they can learn from us.

And that brings me to the fourth part of the ecosystem – the network. While I serve as the Code for America Brigade Captain for the City of Chicago – I'm just one of many civic innovators in a global network. The network allows me to share stories about the work we've done with thousands of civic innovators across the globe.

The real power of this network is it allows us to collaborate and exponentially expand our knowledge base. This is particularly useful for small or newer communities that are just starting out. Instead of having to do things on their own, they can literally go online and ask hundreds of civic innovators a question and get replies within minutes. Because odds are out of the 157 brigades, you are not the only one to have encountered a specific problem. We learn from the network and they learn from us.

This has also helped to scale the civic innovation movement in a very short amount of time. Each new city starts out with the lessons learned from 150 other cities including Code for Japan based right here in Tokyo.

So, I've given you an overview of how we built the civic innovation ecosystem in Chicago. Now, I want to show you what that allows us to do with an example of a civic startup that has come out of Chicago. This here is Rose Arfyie, the founder of mRelief. mRelief is a website that allows users to find out what social services they're eligible for.

Social services in Chicago can be difficult to navigate because the social safety net is so complex. Social services are administered by multiple governmental and non-profit organizations each with their own set of qualifications. There's no one stop shop for social services – you have to apply at each agency separately, which can take a lot of time. It's a big problem.

The start of this effort was when the Mayor's office came to Chi Hack Night to talk about the problem set. Their representatives from the Mayor's office met with Rose. Rose and other participants came up with the idea for mReleif.

With the funding, Rose was able to go full time and hire two additional team members. mRelief used Smart Chicago's Twillio account to enable texting for the app and then went through user testing to ensure the app was working as designed.

Because Smart Chicago works so closely with Code for America, one of the first things that we did was to introduce mRelief to Code for America's health team. Code for America's health focus team has worked extensively with social service issues and with creating apps that text – much like what mRelief is trying to do. This consultation helped to improve mRelief.

Once mRelief launched, they presented their work at OpenGov Hack Night – which Smart Chicago live streamed and blogged about it – helping to share their work nationally.

Later, when Code for America wanted to focus on food stamps for one of their Code Across Challenges – they reached out to mRelief to help run the challenge.

As you can see, the network allowed mRelief to amplify their efforts – and in return, Code for America's health team and Brigade network were able learn from their experience.

mRelief has been a big success in Chicago and Smart Chicago is proud to be awarding them another $15,000 in funding to support their work. (Since the time of the speech, mRelief has also gone on to receive more grants from other non-profits and has provided paid services to the City of Chicago.)

As you can see, the ecosystem helps accelerate projects like mRelief and the network helps to amplify our work.

The History of Civic Technology Events in Chicago

When Chi Hack Night gets visitors from other cities, one of the first things that gets noticed is its size. Chi Hack Night is now consistently reaching attendance numbers of 80 to 120 people and has featured speakers such as Chicago CIO Brenna Berman, WBEZ's Linda Lutton, the University of Chicago's Charlie Catlett and more. Currently there have been 173 hack nights and counting!

Chi Hack Night didn't grow this way overnight. Chi Hack Night is one part of a larger community that's been growing in Chicago as early as April 2009.

It's sometimes easy to think that civic apps are a new phenomena, but there have been people building civic apps long before it became cool. Adrian Holovaty created chicagocrime.org in 2005. (Holovaty would end up founding Everyblock along with Dan O'Neil who would later become the Executive Director of Smart Chicago.)

In 2005, Dan O'Neil also launched CTA Alerts to help riders communicate about issues with the CTA. In 2009, Harper Reed (who would later be the CTO of Obama for America) created an unofficial CTA API.

This post aims to talk a little about the history of civic technology and open governments groups in Chicago and where the movement is going from here.

The 8 Principals of OpenData

While this meeting didn't happen in Chicago, one of the earliest meetings of this movement occurred on December 7th, 2007 in

Sebastopol, California. Carl Malamud and Tim O'Reilly organized the meeting with the goal of establishing principals of open data and with attendees including Lawrence Lessig, Tom Steinberg, Bradley Horowitz and more. The meeting had a significant Chicago presence including Adrian Holovaty, Dan O'Neil, Karl Fogel, and Aaron Swartz.

The meeting laid down the 8 Principals of OpenData, which would later help influence open data policies nationwide.

OpenGov Chicago - April 2007

OpenGov Chicago(-land) meetup group was founded in April of 2007 by Joe Germuska and Daniel O'Neil. The group boasts one of the largest continuities google groups about open government in the country. It's first meeting on Meetup was a social event at Clark Street Ale House. The meetup group was created for citizens who are interested in seeing their federal, state, and local government function more efficiently and responsively. The group is inspired by people who are actively building tools and experimenting with solutions along these lines, like the Sunlight Foundation and GovTrack. The group stated right on the home page that it believes that open source software practices and internet culture provide good examples of how people can work cooperatively on complex problems to produce meaningful results, but you don't have to be a techie to be part of this meetup.

It's worth nothing that OpenGov Chicago stated flat out that you don't have to be a techie to be part of the Meetup. This distinction is important as it helped set the tone that the community here is open.

Independent Government Observers Task Force

One of the first open government events happened in Chicago on August 4th, 2008. The IGOTF Non-Conference brought together CEOs, professors, and nonprofit executives involved in placing case law on the Internet for free access. That meeting was extremely productive in introducing players to each other in a series of cooperative efforts.

Illinois Data Exchange Affiliates

One of the other early incarnations of open government groups was the Illinois Data Exchange Affiliates (IDEA). a voluntary coalition of government agencies and nonprofit organizations working to improve and facilitate public access to public data through web-based XML data transfer. Led by Greg Sanders and Justin Massa, the group met regularly and was eventually merged with the OpenGov Chicago meetup group.

CityCamp 2010 - Chicago

The very first CityCamp was held in Chicago in 2010 at the University of Illinois' Chicago Innovation Center. CityCamp originally emerged from Transparency Camp and the Gov20 Camp. (Here's a google doc featuring the meeting notes from that Transparency Camp that spawned CityCamp.)

At this meeting, Jen Pahlka presented about Code for America - an idea which would grow into a massive international organization with countless people in it's network and a ton of work.

CityCamp itself would also grow into it's own brand with dozens of events happening around the world.

data.cityofchicago.org, Apps for Metro Chicago and Urban Geek Drinks

With the election of Rahm Emanuel as Mayor of Chicago, things began to change quickly. One of the Mayor's first acts was to sign an Open Data Executive Order that set the city on the path to opening up their data. John Tolva was hired as the city's first Chief Technology Office and Brett Goldstein was hired as the city's first Chief Data

Officer. The Department of Innovation and Technology not only published data to the portal, but they hooked up their business systems directly to the portal so that data sets would update automatically.

Brett Goldstein recruited people from the OpenGov Chicago meetup and other technologists to form the "Nerd Herd" - an informal group of people who would meet regularly to talk about technology issues affecting the city.

At the time, there hadn't been many civic apps created in part because there was no data to fuel them. (You had to either get the data through a Freedom of Information Act or just create the data yourself.) As part of the city's strategy, the City partnered with the Metro Chicago Information Center to run the Apps for Metro Chicago Contest to help kickstart projects that used newly opened city, county, and state data.

As part of the yearlong event, MCIC held several hackathons at Google's Chicago headquarters. The content would inject a lot of energy into the community and resulted in the creation of 52 civic apps. The winner, SpotHero, has gone on to become a major Chicago startup and recently just completed a $20 million dollar funding round.

The other big event that was happening at this time was an informal monthly gathering hosted by Justin Massa called Urban Geek Drinks. The venue provided an enormous networking opportunity where people with an interest in civic issues and technology could meet and talk.

Code for America Fellowship, Connect Chicago, and OpenGov Hack Night

In 2012, Chicago became a Code for America Fellowship City with the task of creating an Open311 interface for the City of Chicago with Smart Chicago Collaborative providing funding. At the same time, Smart Chicago was helping to administer the city's Broadband Technology Opportunity Grant and starting to host the Connect Chicago Meetup for people interested in closing the digital divide in Chicago.

Post Apps for Metro Chicago, there continued to be an interest in civic hackathons including two "Idea Hack Chicago" events hosted by Veronica Ludwig, Christopher Whitaker and Josh Kalov in partnership with Code for America as part of their fellowship year.

One of the teams that formed during the very first Google Hackathon for Apps for Metro Chicago contest was Open City - first founded by Paul Baker, Chad Pry, Nick Rougeux, Ryan Briones and Derek Eder. Their entry, Chicagolobbyists.org, was one of the first civic apps to make use of open data and had a great reception when released. The volunteer group had continued to build apps even after the contest, but found it difficult do complete work outside of their 9-5 jobs without experiencing hackathon fatigue. Derek Eder and Juan Pablo-Valez had an idea to create a weekly hack night event as a space to work on their projects.

On March 22nd, Derek Eder and Juan Pablo-Valez hosted the first OpenGov Hack Night at offices of Webitects. A few months later, Code for America would launch the Brigade Program that would strive to harness volunteer energy around civic technology. Christopher Whitaker applied and was accepted as the Chicago Brigade Captain with a plan to support existing work and try and network resources within the community. At the same time, Whitaker joined Smart Chicago as a consultant and was paid to attend hack night and document the movement. At that point, Derek Eder had left Webitects to start his own civic tech development shop Datamade with Smart Chicago being their first client working on projects like the Chicago Health Atlas and Chicago Early Learning.

Over the next few weeks, OpenGov Hack Night would continue to grow and quickly outgrow the offices at Webitects. Through Smart

Chicago's founding membership at 1871, OpenGov Hack Night moved to the IMSA classroom where it would say for the next few years. (Smart Chicago would provide a number of keys to civic technologists over the next few years.)

Several apps that used data to tell stories about the city was featured in the Chicago Architecture Foundation exhibit "City of Big Data" which uses interactive displays to display different aspects of city data.

Over the next few years, OpenGov Hack Night would grow and serve as a model for groups around the country. The space would outgrow even 1871 and move to the offices of Braintree where they are now. As part of the move, they changed their name to Chi Hack Night to reflect caring about more than just open government. Derek Eder and Christopher Whitaker now run chi Hack Night along with a volunteer leadership council with members running their own breakout groups within Chi Hack Night.

Fewer, but more focused hackathons

As regular gatherings of civic technologists became the new normal, it had the effect of reducing the number of hackathons. Instead of having one every weekend, there were fewer but more focused hackathons. Smart Chicago helped to run several of these particularly around National Day of Civic Hacking. Chicago's early success at National Day of Civic Hacking would result in a partnership with organizing organizations like SecondMuse and Code for America to provide training material for newer communities.

Other such events included the Geeks Without Bounds "Everyone Hacks" event at Groupon. The hackathon, co-hosted by Chicago Women in Developers, was specifically targeted towards getting more women into the tech space. After the event, Chi Hack Night saw an increase in the diversity of it's attendees.

Another event that's grown over the years is the Center for Neighborhood Technologies Urban Sustainability Apps competition that connects coders, designers, and developers with community leaders and representatives to solve neighborhood problems.

Other groups

As the community has grown there have been other groups that have formed to help bring together people around the intersection of technology and civic lift. These include the City Data Users Group, Maptime Chicago, the Chicago Data Visualization Group, and more.

Part Two

How To

Civic tech is new. Simple posts that directly document how to do big things were a major part of my beat.

Civic Tech 101

A large part of my work with Smart Chicago involved covering Chi Hack Night (formerly OpenGov Hack Night). By being paid to go to Chi Hack Night it enabled me to take a much greater part in the event. (Before, I would run from my office on Chicago's north side and race downtown to make the event in time.) It also meant being able to be one of the more productive Code for America Brigade Captains since I could wake up in the morning and think only about civic tech things instead of a different day job. This is the talk that I developed when I was starting off as a co-organizer.

When I start out this talk, I always like to ask how many developers are in the room. People who code, know Ruby, have GitHub Accounts. That's because I'm not a coder. I don't know Python, I don't know R and if you let me into your production branch on Github – I will destroy everything.

I came into civic technology from the opposite direction from a lot of my peers.

It was early 2009 and the American economy was in the worst recession in decades. Wall Street was crashing, houses were being foreclosed, and unemployment was skyrocketing. As part of the government's response, the federal government gave money to state governments to hire more bureaucrats to help process unemployment claims. I was hired by the State of Illinois in 2009 to work in a field office.

So, on my first day of work I arrive 30 minutes early. It's early March and so it's still pretty cold in Chicago – and there is still a line of about 20 people out the door. I get to my desk and boot up my computer – and I wait about five minutes because the computer is from 1995 – It's got a big box monitor bigger than my head. I finally pull up the system that I would be using to help my friends and neighbors with their unemployment insurance...and I am greeted by a DOS screen.

This system has green text and it sort of just blinks at me. And I'm floored. It turns out this system was from 1975. The system wasn't a real time system. So, when I did anything (enter in somebody's claim for unemployment insurance, made sure the system sent them a check) – it didn't actually happen until it ran through an overnight process. The only way for me to know for certain if something I did worked was to come in the next morning and print a ream of paper – then go through line...by line...by line... to make sure that nothing rejected. Because if it rejected and nobody fixed it, the claimant would come in 3 days later asking us "Hey! You told me you fixed it! There's no money in my account – now I can't pay rent!"

At the same time, I'm carrying around an iPhone in my pocket.

The iPhone was 10 times more powerful that the system that I had to use to help my neighbors with their unemployment and I'm using the to watch cat videos on my way home. It was incredibly frustrating because to me it should be the opposite. As a front line public servant, I should have the best tools possible to help residents. So, I started getting more into technology as a way to try to find ways to do my job better. I started going to technology events, to hackathons, and looking for ways to get involved.

After graduating with a Master's of Public Administration from DePaul University, I became a consultant for the Smart Chicago Collaborative. Around the same time, I became one of Code for America's first Brigade Captains – and later became the Brigade coordinator for the Midwest region. I also help to co-host the Open-Gov Hack Night in Chicago – one of the largest weekly gatherings of civic technologists in the United States.

Data as Fuel / Civic Hacking in a Nutshell

The next part of the 101 Class is about open data. If you want to customize your talk, you can change the talk to fit your municipality. If your city doesn't have an open data policy, then you can either use Chicago or another city as an example – or talk about you can crowd source open data – but for now the point is to explain what open data is and why it's important.

It used to be that in Chicago, figuring out what the government was doing meant filling out a Freedom of Information Act Request (FOIA). You'd fill the form out, wait a couple days or weeks and then you'd either get a CD with a bunch of excel files on them or worse case scenario – a PDF. And the PDF doesn't do you any good because you can't stuff the PDF inside Excel to run analysis on it.

That was playing the civic hacking game on hard.

In 2011, the City of Chicago implemented an open data policy stating that the city should start opening data in a way that was free for people to use, available to everyone, without any expensive proprietary software. (Following the 8 Principles of Open Data)

To do that, the City launched their data portal at data.cityofchicago.org using the Socrata platform. The site let's us search for data straight from the home page. In Chicago, the City's Department of Innovation and Technology uses an OpenETL Tooklit to link their business systems to the open data portal. So when a city employee does anything – tow a car, inspect a restaurant, fill a pothole – the information is automatically loaded onto the data portal.

But, as great as the data portal is – it's still a giant spreadsheet in the sky. It's not the user-friendliest way to get information.

Take the data about towed cars. That's a lot of license plates to go through. A local civic hacker named Scott Robbin created a site called wasmycartowed.com that's a single sentence. You put in your license plate number and the site will tell you if your car was towed and how to get it back.

That's civic hacking in a nutshell. It's taking data from the government and using it to solve a civic problem or educate the public about a civic issue. There are all kinds of apps that operate on that same principle.

Part Three: Collaborative Civic Hacking with Open Source

Open source and GitHub can get pretty technical sounding and it's a little intimidating since a lot of programmers tend to work from the terminal – which is foreign to most users. In this part of the 101 class, we mainly want to focus on what open source is an why it's important. Don't try to teach the ins and outs of GitHub all at once – that's a lesson for another day. You should refer people to local experts or GitHub's YouTube page for the really technical questions you may get. Instead, try to use an example to show why open source is important.

The other big thing you'll hear civic hacking talk about is the concept of open source and a website called GitHub.

Do you remember test time at school? You weren't allowed to look on anyone else's paper, you weren't allowed to ask questions about how something worked, and you most certainly couldn't copy anyone else's work. This is closed source.

With open source, all of these things are allowed (to a point). Open source software and the GitHub platform allow us to work very collaboratively. My favorite example of this is the Chicago / Boston flu shot app. During one of the first Chicago OpenGov Hack Nights, the Chicago Department of Public Health came to hack night asking about a way to make it easier for people to find flu shots. Tom Kompare, a local web developer, then created the Chicago Flu Shot app that allowed people to use their phones to find flu shots near them. A few weeks after the app launched, the City of Boston declared a public health emergency because the flu had gotten so bad in Boston.

Harlan Weber, the Code for Boston Brigade Captain, shot our group and email asking about the flu shot app. Tom and Code for Boston got connected and the Code for Boston team used a GitHub feature called a "fork" to create their own copy of the flu shot app. They modified it to match the City of Boston and had their app rolled out to the public in less than 36 hours.

But the story doesn't stop there – when the Code for Boston team deployed the app they also made some improvements to the app. They then went back to the Tom and said "Hey, Tom – we made some changes and we think you'll like them. We'd like you to pull our changes into your app." This is called a pull request and we use them to work collaboratively.

Part Four: Break for Questions

With going through both open source and open data, it's a lot of tech to throw at people in such a short about of time. This is a good point to stop and allow for questions.

Part Five: Community organizing as being vital to Civic Tech

A large part of our work involves community organizing and collaboration. The easiest way to explain this is through an example.

Admittedly, we are a geeky bunch. But, if we're geeks buildings things for geeks we're not always solving the problem. In a lot of our projects, we try to partner with either a government agency, non-profit organization, subject matter expert or community activist to help ensure our apps meet the needs and challenges of those in the trenches working to solve civic problems on an everyday basis.

One example of this is schoolcuts.org. In this case, civic hackers at hack night were hearing from Chicago Public School parents who wanted more information about the impending school closing. CPS had released data about each school on the closing list, but the information was scattered across different websites and PDF documents.

The problem was compounded when Chicago Public School announced the list. Not only were certain schools closing, but also some schools were having their locations changed. For parents, this meant that even if their school had been saved from the chopping block – their routes to school would still change. The schoolcuts.org team saw an opportunity to use build a site that clearly displayed the data around the school closing issue in a way that was easy for parents to understand.

Because team members already had connections to concerned parent groups, the team was able to understand the needs of the community and build the site around their needs.

As they built the site, the team kept going back to the parent groups in order to get feedback on the site to make improvements to the site – including translating the site to Spanish.

It's this community organizing aspect that's the driving force behind civic apps that truly make impact.

The final thing that I do is a call to action. This could mean asking people to join your hack night, to break out into working groups, or come to one of your events. For Chicago's Chi Hack Night, I normally encourage people to join our discussion board or head over to one of our breakout groups.

I also use the opportunity to answer any more questions that group may have.

The Schoolcuts.org teams school us on how to build a civic app

----

One of the prime examples of putting people first when developing civic apps was schoolcuts.org.

May 21st, 2013

Since schoolcuts.org first launched two months ago, the team has been working around the clock to add new features and information to the site.

One of the weakness of the civic hacking movement is a tendency to launch a new civic app based on some newly released data set and then never touch the app or the issue again.

It is a rare instance when a civic problem can be solved by one simple app release – particularly when the civic problem is something large and complicated like crime, sustainability, or education.

So while the schoolcuts team launched their app two months ago, they have continued to add more information and features to the site as questions continued around the issue of school closings in Chicago. (Most recently, they've translated the site into Spanish.)

The team presented at Chi Hack Night about how they went about building the app in three separate acts.

Act One: The Problem

The team started with the problem. (Not the data set) In this case, the team was hearing from Chicago Public School parents who wanted more information about the impending school closing. CPS had released data about each school on the closing list, but the information was scattered across different websites and PDF documents.

The problem was compounded when Chicago Public School announced the list. Not only were certain schools closing, but some schools were having their locations changed. For parents, this meant that even if their schools weren't closing – their routes to school would still change.

Additionally, the school attendance boundaries were not matching what Chicago Public Schools had designated to be the receiving schools. In effect, your oldest child may be going to a different elementary school that your youngest when they reach school age.

Further, the criteria used to closed schools used an additional value added system that further confused parents and community members.

Act Two: Opportunity

The schoolcuts.org team saw an opportunity to use build a site that clearly displayed the data around the school closing issue in a way that was easy for parents to understand.

Because team members already had connections to concerned parent groups, the team was able to understand the needs of the community and build the site around their needs.

Act Three: Solutions and Challenges

To help parents, the team decided to try and ensure that the user experience was very localized to their school.

The site not only displays the data around school closings, but also does a good job of explaining what the terms mean.

Since their launch, the team has continued to add features such as a simple way to compare closing schools with receiving schools and has translated the site into Spanish.

The schoolcuts team shows the right way to approach civic apps. They address a community concern using open data and educate the public about the issue.

How to run a civic hackathon

As part of my work with the Smart Chicago I've run several civic hackathons including several National Day of Civic Hacking events in Chicago. Starting in 2014, I wrote a post on how I organize and set up these events. I also wanted to go over tips and hazards to look out for when organizing these events.

First off, hackathons are good for some things and not for others...

There are some common critiques of hackathons that result from a misunderstanding on what civic hackathons are good for. We also want to differentiate between regular industry 'hackathons' that can double as marketing events and civic hackathons that have a much different purpose.

Hackathons are a tool for community organizing or organizing around a specific problem set.

Hackathons are not useful for:

- Building fully finished apps: The "Oh, let's have a hackathon to build an app to solve all our problems!" type of hackathon generally flops because a weekend isn't enough time to build a finished fully capable product.
- Getting people to use YOUR product: The "Oh, let's have a hackathon and get people to use our platform to get more customers!" generally isn't a good idea. Even with a prize for the best use of your tool, you aren't really hosting a civic hackathon. You're hosting a marketing event. Not that there's anything wrong with that, but when you do this you are missing the point of hosting the civic hackathon in the first place. (Which should be spur civic innovation to solve social problems.)

- Building a very specific app that meets your very specific needs: The "We really want this, let's order a bunch of pizza, hold a hackathon and have them do something!" is counter-productive. If you have a scope of work and know what you want, just hire somebody.

On the other hand, civic hackathons are useful if you want to:

- Organize people around specific problem set and brainstorming ideas: "We want to make our city sustainable. Let's get the environmental affairs people together with the technologists " type of hackathon.
- If you have a particular problem and a particular set of resources you can throw at it, a hackathon can help brainstorm ideas about how to address the problem. Keep in mind, this hackathon only works if you have subject matter experts *and* the technologists talking together.
- Organize to recruit people into the civic innovation community: The "there are not enough women/non-profits/journalists at our hack nights" type of hackathon. Hackathons can be great recruiting tools to get more people into your community. Granted, these types of hackathons are built for beginners and have more of a training element to them.
- Test out a new civic dataset or API: The "Ok, we finally got this dataset out.. now what can we do with it" type of hackathon. Fair warning: the data in it of itself won't be enough. You'll need to have subject matter experts on hand to help explain what the data is actually showing. For example, if you've released a new food inspection data set you should have somebody from the health department on hand to explain the details. (People are not good guessers). These types of events are good for seeing the types of projects that can be done with your new data set or API.

We want to start off planning for the hackathon by asking, "What is it you want to accomplish?" That also leads us to a major point in having a successful hackathon.

Having a plan to carry on the work after the event.

While civic hackathons are a great organizing tool, they don't always result in fully finished and tested civic apps. For that, other efforts need to be made to carry the work forward.

When thinking about your event, you should keep in mind having a plan for carrying the work forward. Does this mean starting your own weekly hack night? (So you can continue to work on apps started at the event a little bit at a time.) Does this mean finding grant funding to support the development of larger apps? (Using the hackathon to show the potential of these apps.) Don't discount the power of small wins – particularly when your local civic innovation community is just starting out.

Hackathons can be a great spark to larger efforts. With that pre-text in mind, here's the Smart Chicago approach to hackathons.

First, start with a problem:

Civic hackathons work best when focused on a specific problem set. The best way to highlight these problems is to bring in a subject matter expert who deals with the problem on an everyday basis. What to do a hackathon about crime? Better call the police department. Want to do something around housing? Recruit somebody from the local housing authority or somebody who runs a shelter. When trying to think about the needs and challenge of the front line, there's no substitute for somebody who actually works in the front line to speak at your event.

You also shouldn't wait until the day of the event to provide your participants with resources. By providing a resource list ahead of time (like this one we put together for the Chicago Police Department Safer Communities Hackathon) it helps get participants better prepare for the event. Hackathons may have challenge components, but that doesn't mean everyone absolutely has to start from scratch.

Getting an agenda together

Your agenda will vary depending on the goal of your hackathon and it's something that you should determine early on in your planning. There's a temptation to have hackathons become two-day sprints with participants getting little to no sleep. However, if the goal of the hackathon is to develop prototypes versus fully working apps – you may not need really need to make a coffee run at 2 in the morning.

The formats for our event tend to follow this general flow depending on what we're trying to accomplish:

- 7:00am Event organizers arrive to help set up.
- 7:30am Food/Coffee/Water arrives
- 8:00am Breakfast
- 8:30am Welcome ("Hi, welcome! Here's the wifi password, the bathrooms are over there.")
- 8:45am Education ("Here's what we've done so far, here are some resources")
- 9:00am Problem Set ("Here's the problem I'm facing in the day-to-day")
- 10:00am Q & A ("Would this be helpful?" "How does this activity work?")
- 10:15am Break out to form teams and hack!
- 12:00pm Lunch served

4:00pm Hacking ends, judging begins (if applicable)

4:30pm Project winners announced (If applicable)

Variations on this include having the talking portion of the event happen on a Friday night. (This is particularly helpful if you're trying to have working prototypes by the end of the event.) This allows for a full day (or full two days) of hack time.

If we're doing more of an unconference event, the format will go something like this:

- 7:00 am Event organizers arrive to help set up.
- 7:30 am Food/Coffee/Water arrives
- 8:00 am Breakfast
- 8:30 am Welcome ("Hi, welcome! Here's the wifi password, the bathrooms are over there.")
- 8:45 am unConference 101 ("Here's how unconferences work")
- 9:00 am Icebreaker ("Everyone is going to form a line based on a statement. If you agree head to the right, disagree head to the left. OK, the weather is too cold!")
- 9:30 am Throw session ideas on the board, vote, assign rooms
- 10:00 am First Session
- 11:00 am Second Session
- 12:00 pm Lunch served (Throw out and vote on afternoon sessions)
- 1:00 pm Third Session
- 2:00 pm Forth Session
- 3:00 pm Group Returns & Report backs (Here's what we talked about)

These are just general examples and in the course of running the event you may find yourself running a bit behind. We always assume that things are going to run over by about 15 mins and just prepare accordingly. (We sometimes use the lunch hour to make up for time.)

Next, make sure you recruit a diverse set of participants

Once you have an agenda, it's time to Recruit! Recruit! Recruit! People can get pretty busy and people can get even busier during the summer. In order to get people to your event you'll need to start early. You can use your local organization's email lists, social media accounts, contacts with local universities, companies, and news media to help spread the word.

The other thing you'll want to make sure of is that you recruit a diverse pool of participants. Reaching out to groups like the Chicago Women in Developers can be super-useful in doing this. Additionally, you'll also want to make sure you have a diverse skill set at your event.

The best teams at hackathons are comprised of a mix of skillsets including developers, user interface designers, data scientists, and subject matter experts.

There are a few strategies you can use to try to get a diverse group of people to your event including reaching out to Meetup groups (like the Northern Illinois PHP group or the Chicago Interactive Design & Development Group), contacting web development schools like DevBootcamp, and talking to local universities.

There's a few ways that you can do registration for the event. For our groups, we like to use Meetup since it allows us to host discussions on the event page. It also helps to carry the movement on afterwards since you'll have people joining your meetup group when they register for the event.

Venue and Logistics

Venues can make or break an event. A good venue will help you to ensure that the event runs smoothly, while a bad one will cause you to bounce around like a pinball trying to fix last minute everythings.

Co-working spaces like Chicago's 1871 tend to make great venues. They're designed to handle a large group of people all getting on the internet at the same time. They're large enough so that people can break out into smaller groups and still be able to hear other. Many co-working spaces already come with their own audio-visual equipment. And, most importantly, co-working spaces often have staff on hand to help out.

For unconferences, large offices or colleges are also good options since you'll need a lot of rooms to be able to run breakouts in.

When we look for venues to run our events here's a list of things we look for:

- A location easy to get to, ideally near mass transit.
- A wifi network that can handle a hundred people at once. (Not always easy)
- The venue has either a lot of places to plug in or the ability to place more power strips out.
- A venue that has projectors and screens.
- A staff member that can be there during the event.

A venue can't provide anything. Most of the time, this just means food. However, this can include other random items such as:

- Dongles/Adapters (Because nobody has a laptop that will connect to your projector. You need one that links to VGA, Mac to HDMI, Mac to VGA, and so forth and so on.)
- Name Badges
- Snacks, Water, and Pop/Soda (remember to bring the healthy stuff)
- Power strips
- Post-it notes
- White Boards / Paper Easels
- Microphones / PA system
- Projector, Screen

- USB Drives
- Swag (laptop stickers, t-shirts)
- Directional signage (Where's the bathroom?)

Venues can also be one of the bigger expenses, which brings us to our next topic.

Sponsorships

There are a number of costs that are associated with running a hackathon including:

- Food (We like deep dish)
- Supplies (Pens, markers, paper boards, power strips)
- Venue
- Your time (Even if you volunteer, you're still donating your time)

A variety of sponsors can help defray the cost of running your event. If you're a Code for America Brigade, you're already getting some support. Chicago's Chi Hack Night actually has a rotating list of sponsors including Smart Chicago, Thoughtworks, Code for America, and DataMade. You can also get in-kind donations such as the use of the venue, food, and use of software platforms and hosted space. (Smart Chicago Collaborative is part of the Chicago Community Trust and is funded in large part by the John D. and Catherine S. MacArthur Foundation. We *highly* recommend reaching out to your community foundation.)

Punch List

If you prepared everything right, then the actual day of the event can be easier then doing the prep work for it. Here's our punch list of things to do before, during and after the event.

- Venue walkthrough (everything look ready to go?)
- Reminders to attendees?
- Security has list of names? Including yours?
- Food ordered?
- Social Media Set-Up
- Speakers slides? (Do you know how to pronounce their names?)
- If you're speaking, did you rehearse?
- If you wanted press, did you talk to press?
- Did you do a final check-in with the other organizers?
- Sleep. (Seriously, go to bed.)

Day of:

- Signs out?
- A/V Up?
- Name tags set out?
- Chairs/Tables how you want them?
- Breakfast set up?
- People know where to go?
- Are you taking pictures?

After:

- Thank you notes
- Blog post about the event
- Follow-up to ensure work continues

Hackathons, particularly large ones, can take a lot of work. However, when done right they can be huge spark that kicks off civic innovation efforts in your city.

It should be noted that not all civic tech events have to be hackathons in nature. They can be training camps, town halls, or unconferences. The same lessons apply however.

How to run a Chi Hack Night-style event

ChiHackNight has been running in Chicago for almost three years. It has grown from a handful of technologists a week to one of the largest gatherings of civic technologists in the country with a diverse group of attendees from all different fields. What distinguishes ChiHackNight from other events is the emphasis on being a place to learn and to provide a space work on projects. The hosts do very little project management, and it's primarily the attendees that drive hack night projects. Below is a post we wrote up describing a Chi Hack Night format in summer 2014.

July 24th, 2014

Over the past few weeks, there's been a new format at Chicago's Chi Hack Night in an effort to bring more leaders into the management of hack night.

Before and After

Before, the hack nights had a consistent agenda.

- 6:00pm – Food
- 6:15pm – Introductions
- 6:30pm – Announcements & Presentations
- 7:00pm Hacking / Civic Hacking 101 (From 7:00pm – 7:20ish)

This worked pretty well, but it wasn't always easy to do the connection between arriving at hack night to working on a project. This particularly became true once our attendance became larger. Additionally, the hack night wasn't set up in a way where people could step up to be leaders.

The format has now changed in the following ways:

Leadership Council

Previously, just three people organized the hack nights. Now, a volunteer leadership committee that meets regularly runs the hack nights and anyone can join. The general guidelines for the meeting are based on the ones that FreeGeekChicago uses.

We want everyone to have a voice. At the same time, we don't want discussions to become unfocused or endless. So we try to follow some guidelines:

- We try to keep the meeting length less than one hour.
- There is a 10-minute time limit to all topic discussions. If there is no solution after 10 minutes, we poll the room to decide whether to discuss the item further or table it until the next meeting.
- The meeting notes are then posted in the ChiHackNight Google group.

The Breakouts

The other change is that there are now multiple breakout groups focusing on different subject areas. This is being done as a way to help facilitate projects and work being done in each area – even if there's not a particular app being worked on. The breakout groups are divided into two groups – learning groups and working groups. The groups are also flexible, so that if people have an interest in starting a new one they can.

Learning Groups:

- Civic Hacking 101: An orientation into civic technology and the event
- Code Clinic: A place where people who are learning to code can ask questions.

- Mob Programming: Hands-on learning! This group shares a single screen, keyboard and mouse and use a timer to rotate control among participants to complete a shared exercise.
- Mapping all the data we can find: Hands on workshop for getting your data mapped and analyzed in the cloud.
- Dolphin Tank: Develop project ideas, connect to talent and resources, or find new projects to join in a safe, friendly space with your peers.

Working Groups:

- Access to Justice
- Education
- Environment
- Pension Reform
- Social Service Delivery
- Transportation

How to Blog

One of the most important roles I've played at Smart Chicago is to serve as an archivist to the movement. I've currently published more than 330 posts. Here's a post I put together on our system as a teaching tool for some Smart Chicago interns.

It always starts with a nice Creative Commons photo

Flickr is a fantastic tool for finding pictures. We either use Flickr's image embed tool to copy the code directly into WordPress or we just use Wordpress' native tool. If we're showcasing a specific website or tool we'll use a Chrome plugin called "Awesome Screenshots" to take a picture of our screen.

WordPress is wonderful and has several features that make blogging easy. One of these features the "More" feature. One some Wordpress templates, posts can get long and make it hard to scroll through all the pages. After we introduce subject of the post, we'll include the "More" tag to the text to create the post's 'fold' so that our home page doesn't get too cluttered.

Once we've introduced the topic, we'll talk more about the problem we're trying to solve.

About the problem

We start off giving examples and screenshots of the issue we're talking about. We also want to put the problem into context. So, if we're dealing with something just applicable to Chicago or in the context of a certain time-frame we want to bring that up.

Data

Because we deal a lot with open data, we always want to point out when data exists already. Chicago's civic technology community mainly gets data from three places.

- The City of Chicago data portal at data.cityofchicago.org
- The County Data Portal at datacatalog.cookcountyil.gov/
- A Freedom of Information Act Request

WordPress has numbered lists and bullets that we use a lot too. You'll also notice that on our blog we tend to link out to different websites. We do this because we want to show where we got our information and because we want readers to be able to go straight to the source.

The Solution

The bulk of our post will be about talking about the actual solution. If it's an app, we'll run down the parts of the website. However, it's no substitute for an actual working demo. For this we do one of two things.

1) We use a program called Screencast-o-matic to do the demo. Screencast records both the camera and the screen. We then upload this into YouTube and use their embed feature to post directly on our site.
2) If we're covering a presentation at hack night – then we'll use the video we recorded there and upload that.

When we go to Hack Night – we've used both Google's Hangouts on Air feature and a program called Wirecast. When we do that, it uploads the video directly to YouTube. We then download that video and put it into iMovie. At that point, we'll slice the video into shorter parts and clean up the audio. That gets uploaded to YouTube and embedded in the post. To film, we use an HD webcam and a tripod. We also

How it came about

Many of the solutions that come out of Chicago's civic innovation community have more to do with the community organizing aspect of the work more than the technology side. We always want to stress this and talk about how everything came together. All of the solutions we talk about have to replicable – and for that it takes people working together.

Call to Action – Getting more information

The last thing we want to do is to give the reader a call to action. This could mean registering for our event (like the Chicago School of Data conference in September) or filling out a form (we use wufoo) or clicking a link so that you can get more information.

When it's done

When we're done writing, we'll submit it to our Executive Director Dan O'Neil for review. We almost always have another set of eyes look at a post because proofreading and editing are important.

When we do publish, we tend to do three things after.

1) We use a WordPress plug-in that immediately posts to our Facebook page.
2) We'll tweet it out being sure to mention any partners or people we covered.
3) If it's a topic related to one of our partners (like Code for America) we'll sometimes cross-post it.

Live tweeting and using Twitter to boost your event

I use social media significantly to help spread the word at our events as well as to share what's happening at different civic events throughout Chicago.

I'm going to go over a few tips and tricks for using social media to boost your event.

You'll usually have two goals with social media. The first is to get people to attend your event. The second is to add followers so that when you have future events or news you can spread it more easily.

Twitter

I use Twitter when we're covering live events. Twitter's ability to post rapid real time updates makes it perfect for things like this.

My strategy for events is to write up a post advertising the event. If we know the hashtag already, we'll start using that when we tweet the event out.

I use the hashtag so that people can start following other accounts that are also using the hashtag. This also lets our followers know there's an event going on and that's the hashtag we're using.

I'll also retweet other accounts that are using the hashtag. Sharing is caring.

Just before the event starts, I try and ensure that we're following all the speakers and organizers. Once the event starts, I'll begin livetweeting.

When I'm live tweeting, I normally have an aftermarket tool like Hootsuite or Tweetdeck so that I can tweet from both my personal account and my organization account. I'll have one column set up just for the hashtag so I can quickly retweet relevant tweets.

I'll also have a phone in my hand and logged into Twitter as my organization account. Tweets with pictures tend to get more engagement. I normally do a photo when the event start, when a new speaker comes on stage, or there's a particularly interesting quote. Whenever we take any photo or mention something somebody said, we almost always tag the person if they're on Twitter.

For national events like Code Across and National Day of Civic Hacking, I'll also use the national hashtag as well as mention the @codeforamerica and @civichackingday account. When an account with a large following retweets your tweet, it can significantly amplify your message.

After the event, I'll use the twitter stream as notes for when we blog about the event.

A note about trending and gaining followers

A lot of times, particularly on television networks, you'll see people encouraged to 'make something trend' as if it's a game you can win. We do not advise this.

Trending doesn't measure popularity - it measures velocity. It tries to show what 'new' topic people are 'now' talking about. Once a lot of people have started talking about something, it loses its status as a trending topic. That's also why you don't see Justin Beiber, One Direction, or any of the other boy bands trending all the time. That's also why during television shows; you'll see the networks make up hashtags on the fly. They know that the odds of trending go way down over time. It's easier to get a new hashtag trending rather than an old one. If they tell people to tweet at a hashtag for an upcoming episode, they lose the needed velocity to make something trend. Once it starts trending, it's tough to get it to stay trending because it relies on 'new' people tweeting the hashtag. No points are awarded for the same people tweeting the same thing a bunch of times.

Which brings us to, why try to get something to trend in the first place?

For television shows, it's about advertising. They want people who are just cruising twitter to see the trend and think, "Oh wow, a lot of people are watching this – maybe I should tune in."

For this community, it's not important. Your goal isn't to trend, but to build your audience. This is particularly true if you're running the account for your civic tech group. You want people to start following you and learning what you're about and how you can get involved. That means that buying followers won't do you any good.

Getting Twitter followers takes time and consistently producing content worth tuning in for.

Analytics for Twitter

It used to be that you had to pay to get stats for Twitter. That isn't true anymore. You can access Twitter analytics for free by going to https://analytics.twitter.com and see how you're doing.

Tech Tools for Civic Organizations

One of the problems that we sometimes encounter in the technology space is that we say things like, "Oh, just use this piece of software that I assume you know about." Technology organizations tend to have a "stack" of tools they work with and assume everyone else has heard of.

It's the digital equivalent of watching a home improvement show and they get out a circular saw. The show says, "Oh, just make a few cuts here."

The problem is that it assume the viewer even has a circular saw – and the clamps, goggles, saw horses, the working space, and the know how to actually use the expensive piece of hardware that can seriously hurt you if you don't use it correctly.

Which is kind of crappy.

The good news is that there are a number of tools that are easy to use and won't break the bank. We tend to favor lightweight tools because they're easy to use, not expensive and we can use them in solutions that are repeatable.

WordPress

WordPress is the backbone of our digital communication strategy. It's what runs not only our website, but the website of many other organizations as well.

WordPress is easy to set up and with a few additional changes you can have the site point to your own domain name. There are two options to do this. The first is that you can use wordpress.org to set up a custom install on your server. However, we recommend just using WordPress.com (which does all the setup on the their side.) Initially, you'll have a website.wordpress.com site. You can then pay to upgrade to have the blog point to your own homepage site once you buy a domain name.

Once your site is set up, you can choose a template for your website. Feel free to experiment to find one that fits your needs. WordPress also has a number of plugins that can be used to improve your site. For example, there's a plugin to show tweets from your social media accounts.

Google Drive

Google Drive is a set of office tools where the documents live on the web rather than your hard drive. It includes Google Docs (Word), Sheets (Excel), Slides (Powerpoint), and a few other applications. Having documents that live online means that you can access them from anywhere including your phone.

However, the real reason that we use Google Drive is that it makes it super easy to share documents and edit collaboratively. Instead of emailing revisions of the same doc twenty different times and having to rename the different versions, you can simply share one link to the team. You can also post comments and have running conversations using the embedded chat feature. During OpenGov Chicago Meetups and Chi Hack Night, we tweet out a link to a doc anyone can edit and take collaborative meeting notes.

Google Drive is also pretty easy to use if you're already used to working in Microsoft Office.

Slack

Slack is an internal chat room. It's a more modern version of IRC with many more additional features including being able to integrate with everything from Google Drive to social media channels. Slack allows you to add different channels in addition to the standard "General" and "Random." When we use Slack, we have a separate channel for all of our projects. Slack also has a powerful search feature that can be useful when trying to remember something that the group was talking about from weeks ago. If your organization ends up sending a lot of small two sentence emails, this may help cut down on that.

Slack can just sit there in the background while you work. If you need to get somebody's attention, you can mention them by adding a @ to their username (like Twitter) and it'll send them a notification.

Slack also works well on mobile devices. While Slack has an app for Mac and Windows, you can also just use the browser.

Slack also allows for a lot of customization. It's still a fairly new product, so the company is also still adding features.

Mailchimp

Email is still one of the biggest ways that organizations communicate with their communities. Mailchimp helps organizations by first helping to craft well designed eye-catching emails, but also by helping organizations manage email campaigns. You can pick customized lists of recipients, monitor opens/reads, and even conduct A/B testing of different email campaigns.

Wufoo

At its core, Wufoo is an online webform builder. You can login and create a form in a matter of minutes, then embed it on your blog or just link to it. It's a very simple way to get resident feedback, run a contact us page, or run surveys. When people complete the survey, you can have it send you an email to notify you. Wufoo can also export responses in a CSV file too.

What really makes Wufoo our preferred tool is that it has an API that we can plug into other apps. When we run the Civic User Testing Group, we use a combination of Wufoo and Mailchimp to manage our signups.

Textizen

Textizen is a survey tool that uses SMS messaging to get people's feedback. You can create a survey within a few minutes, but the team at Textizen can help you craft a survey so that you get the best results possible.

The way textizen works is that you set up a survey and it assigns you a phone number. You then create signage that lists the phone number and the first question.

It then will send text survey questions back and forth. As the owner, you can see responses in real time and then export them to whatever format you need.

Smart Chicago Collaborative offers the use of Textizen for free to any civic developer in the Chicago area.

Meetup

Meetup is a tool to help run meetings. We use Meetup extensively to run OpenGov Chicago and Connect Chicago. Meetup is more than just an RSVP system. Once somebody joins your Meetup, they're considered part of the group. This means that you can communicate with group members anytime – such as when you host your next event.

Meetup also lets people posts messages to the group during events so that you can keep a running conversation about what's going on. (We usually use it to post links to the meeting notes or livestreams).

Twilio

We love apps that text. As an organization that cares about digital access, we're all too aware that not everyone has access to the Internet. So, allowing our apps to text ensures that everyone can use it.

Twilio is the equivalent of giving your app a cell phone. When you sign up for Twilio, Twilio will assign you a phone number to use. You can then use the phone number to send and receive text messages. If you've ever order pizza or an online delivery and gotten a text message right after, then you already have an idea of how this works. Certain actions will cause your app to send out a text.

One of our favorite uses of this is the humble CTA bus tracker app. If you go to any bus stop sign in the city, you'll see a short code and a number. You text the number on the sign and it tells you when the bus is coming.

While you do need a developer to use Twilio, we wanted to include it because of the utility of texting. Twilio's API is easy to use if you're a developer with some experience using APIs.

The toolset we use changes as needs and new tools come out, but we always try to stick with tools that are easy to use and are not overly expensive.

Collaborative Project Management

Collaboration is at the heart of the Smart Chicago Collaborative and is essential to achieving the goals of the civic technology movement. The hard problems that need to be solved cannot be solved in isolation.

There is an art to collaboration. Being in a collaboration means that you've agreed that your partner or partnering organization is already highly capable at what they do. It means that you've agreed upon a common goal and a plan of action to achieve that goal. Being in a collaboration means that you've opened up the lines of communication for the duration of the project.

Working collaboratively isn't always easy – particularly when the project involves multiple partners or complex problems. Things can get exponentially more complex each time you add a moving part. Here are some thoughts on how I've approached collaborative project management in my consulting practice and in my work at Smart Chicago.

Building a plan

One of the most difficult things about project management the natural tension between sticking to a plan and the need to be able to adjust the plan if something doesn't go according to plan. This is important because things never go perfectly according to plan. The planning process is important for working on collaborative projects because you're going to have multiple parties running activities that are independent of each other in order to accomplish the goals of the project. If there's not a well-defined plan, the project could steer off course. Here's our workflow for planning projects.

Step One: Gut Checks

Define the goal, ensure the activity matches the goal, and ensure the goal is in line with your organization's mission statement

The first part of taking on a project is establishing the goal and ensuring it the activity being proposed will get the results you want. You're essentially asking the question, "At the end of the day, what do we want to be the result from this project?"

This is a different than saying, "We want to run a hackathon." If the goal is to have an app that gets resident feedback, then you don't need a hackathon. There are existing products that do that. If the intended results are, "We want to engage the civic tech community around environmental data we just released", then a hackathon would accomplish this. It's perfectly OK to have one idea about how to get something done and then change tactics once you review it. Once you determine that the goal will achieve the objective you can move on.

The other thing that you will need to check for is that the project goals are in line with your organization's mission. Would getting these results also help your organization achieve its mission? Is your organization's mission to help with FOIA requests? Does taking the time to write, print, and distribute a newsletter help you achieve this? Is your organization in the business of creating technology tools to help store data? Does running a event help with this? If the answer to the question is no, then it may be advisable to say no to a project. It isn't always easy to say no, but performing this 'gut check' before you begin is a good way to ensure the resources of your organization are being spent wisely.

Once you perform these 'gut checks' and everything checks out, you'll be ready to begin constructing the plan.

Step Two: Work backwards from the goal and make a list of everything you'll need

Think about the activity that you're setting out to do. For example, you want to host a hackathon helping an environmental agency engage with the developer community about new environmental data. What does that look like?

At the end of the event, you'd have people presenting on prototype projects to all event participants after a day exploring the data sets. (Needs: Venue, microphones, projectors)

You would then work backwards through the day and mark down items that come up. For example, before the presentations the hackathon would have teams working with the data and coming up with prototypes. (Needs: Wifi, power strips, things to write down brainstorming notes on, lunch, breakfast.) They'd also need an orientation into the data and the problem set. (Speakers for the morning, links to different data sets)

You would continue to work backwards until you get to your current point in time. For our hackathon example, this means going over getting people to come to the event. (Needs: RSVP page, blog posts, invites, social media outreach).

Working backwards helps ensure that you don't forget anything in the planning process. At the end of the exercise, you'd have a list of items that you would need to have ready as well as a draft of a possible timeline.

Step Three: Take inventory of resources and identify gaps

The next thing you want to do is to ensure you have the right resources to achieve the goal of the project. This often means taking stock of the resources all the organizations have at their disposal and matching them with the needs of the project. This is often the very reason to work on projects collaboratively. It allows two organizations with different strengths and resources to combine them into achieving a goal. There might be times when you check the list of needs and resources and discover that you're lacking in an area. At that point, you have the option of hiring a vendor or reaching out to an organization with resources you're looking for and partnering up.

For our hackathon example, I have lots of experience running hackathons. However, Smart Chicago doesn't have any subject matter expertise working on environmental issues. This would be a gap to fill.

As you work through matching needs with available resources, you'll also notice a list of items that nobody can really provide – but can be easily purchased or ordered through a vendor. (For our hackathon example, this is food and flip charts.) This help with our next step.

Step Four: Determining a budget and harmonizing the plan with the budget

Once you determine the project needs and what resources you have on hand, it becomes easier to establish a budget. Depending on your project, you can price out some things pretty easily. (For our hackathon example, it's a simple process to start calling up different catering companies and getting estimates for food.)

If your project involves hiring a developer, designer, consultant or other freelance position, then additional steps may be involved. You may need to put out a Request for Proposals (RFP) that include the deliverables you need and how soon you need it. We'll get into the details of writing good RFPs in a later chapter. For right now, understand that it may take some time for vendors to put together responses to an RFP depending on what you're requesting.

A final thing you should account for when making a budget is taking into account the cost of your own time. Even if you're volunteering, you should account for the hours you spend working on the project.

As the costs of the project begin to shape up, you'll get a sense of if your organization has the resources to pull off the project or if the organization needs to obtain outside funding through grants or sponsorship. Alternatively, your organization may have been awarded a grant to fund a project – and you have to ensure that your project plans fits into the budget.

Step Five: Writing out Scope of Works and Memo of Understanding (MOU)

At this point, you should have a pretty good idea of what you want to do, what it's going to cost, who is doing what, and a general timeframe. Now it's time to get those thoughts down on paper. The most common ways to do this for collaborative projects are Scopes of Works and Memos of Understanding (MOUs).

Scope of Work

A Scope of Work is a document that's written in the planning stages of a project. They're often written in the context of one person or organization hiring another. If there's more than one partner that'll be doing the work, there'll be more than one scope of work. Scopes of Work lay out what work is to be done, how quickly it's to be done, what report backs need to be done in connection with the work and (in the case of a vendor/client relationship) what the estimated compensation will be. Good scopes of work are flexible enough to allow the person or group doing the work to best decide how it gets done, but strict enough to say what the result will be. (Good example, I need a website that helps people find public computing locations. Bad Example: I want a website written in Python that let's people find public computing locations by searching a Google Fusion table.)

Scopes of work can either be written by the client or the vendor as a response to an request from a client. It doesn't matter so much who starts the process, but rather that both parties agree on the final scope of work. Even if both parties have a pretty good understanding of what needs to be done, going through the process helps to make sure everything is crystal clear. It also helps set limits on the work to be done in case the project turns out to be much more complex than first realized.

Once the scope of work is finalized, they can often be turned into contracts that organizations can use to pay vendors or consultants.

Memo of Understanding

A Memo of Understanding (MOU) is similar to a scope of work on that it sets up what work is to be done, who is doing it, the expected timeframe, and what report backs are needed. The main difference between a Scope of Work and a Memo of Understanding is that a MOU Is used between two parties when no money is being exchanged.

The MOU is important because it sets up exactly who is doing what at the very beginning. If there's ever a disagreement on who is supposed to be doing what, both parties can refer to the MOU.

MOUs and Scopes of Work are essential to collaboration because they establish a clear understanding of the project. Once these are signed and agreed upon – or turned into contracts if needed – you can then start on the project.

Step Six: Iterative Process: Checking in on progress and making adjustments

One of the drawbacks to technology projects is that they take a lot of time and effort to create. Given the costs of hiring developers, technology projects can also get expensive very quickly if projects are not managed correctly. Additionally, if project managers wait until something has been delivered it and it comes out wrong, it can take a significant amount of time to correct.

An iterative process favors a short cycles of work, check-in, and adjustments. During the check-ins, the expectation is not to have made progress in leaps and bounds – but rather to have made smaller updates. Because the progress is in small increments, it's much easier to make adjustments than it after a team has spent several weeks working on a product.

It also means that you can catch blockers early so that they can be resolved quickly. Does the team need more resources? An extra team member to get it done? Is one of the vendors not up to the task and need replacing?

There will be some cases when the blocker is large enough that the time and effort it would take to resolve it outweighs the cost of doing something else entirely. For example, your team is working on an app to analyze a dataset that was received from a Freedom of Information Act Request. It turns out that the data is far more dirty than first anticipated with missing data, misspelled entries, and obvious typographical errors. The amount of time that would be needed to clean it up far exceeds the timeline first established. The question that you would face as a project manager is do you continue the course and accept a longer timeline? Do you narrow the scope of the project and drop the data that is the most difficult to clean up? Or do you add additional resources to help with cleaning up the data? There's no specific right answer, but these are the kinds of challenges that may pop up as managing collaborative projects. (For reference, when Smart Chicago was faced with a similar problem we narrowed the scope.)

In addition to discovering blockers early on, working iteratively also allows for testing of products in front of real users. When the team has something that's somewhat close to the final product, they can have the project undergo user testing to ensure it's going to work for the user like it's supposed to. If it doesn't, then rather being a failed project, the team runs through another iterative cycle and makes improvements before launching.

Check-ins with partners

You should also schedule regular check-ins with partner organizations. This lets everyone know where everyone is at, what the current blockers are, and if there's any adjustments that need to be made to the plan. By keeping everyone well informed, it also helps work more collaboratively. Surprises are great for birthday parties, but not when managing collaborative projects.

Step Seven: Once it's done

Once you've completed the activity that you had planned out before, it's time to let people know about what you did. Whether it's a new app launch, successful hackathon event, or a new guide on how to run civic technology projects – you should tell people about what you did. You can do this through blog posts, social media, or email campaigns.

It's important to tell people about the project so that people can learn from your actions. The more people learn about your actions, the more it advances the field.

Health and Human Services Primer

In 2015 for National Day of Civic Hacking, I wrote up a number of primers on different civic issues to help people get a better understanding of the issues as they start working on projects.

Healthcare and the social services that are often connected to it is an extremely complicated and expensive issue. According to the Kaiser Foundation, the United States spent about $2 trillion dollars on health care.

In addition to regular healthcare, state and local governments spend a tremendous amount of funds on social services. The effects of the recession, pension crises in multiple states, and cuts from Congress have caused state and local governments to make drastic cuts to social services. This happened at exactly the time that more people required social services putting significant strain on the social safety net.

Aside from the big picture, the experience of those receiving social service is an innately human one – and an experience that most Americans don't have. Most Americans at some point have to go to the DMV and the experience is often portrayed as downright in pop culture as downright terrible. The experience of being on social services often is worse – not in just the big picture sense, but in small ways. The social safety net is managed by a multitude of government agencies and nonprofit organizations.

Technology projects, such as mRelief, center on health and human services often try and help residents better navigate and understand the resources available to them.

Data Resources

Data.gov/health

The federal data portal contains over 800 datasets on health. We've highlighted some key ones:

- Hospital Charge Data: Data are being released that show significant variation across the country and within communities in what providers charge for common services. These data include information comparing the charges for the 100 most common inpatient services and 30 common outpatient services. Providers determine what they will charge for items and services provided to patients and these charges are the amount the provider's bills for an item or service.
- Community Health Status Indicators (CHSI): to combat obesity, heart disease, and cancer are major components of the Community Health Data Initiative.
- CDC Cancer Statistics: The United States Cancer Statistics (USCS) online databases in WONDER provide cancer incidence and mortality data for the United States for the years since 1999, by year, state and metropolitan areas (MSA), age group, race, ethnicity, gender, childhood cancer classifications and cancer site.

County Government Data

Many county governments administer their own health and human service systems – some of which release this data to the public.

Cook County, IL (https://datacatalog.cookcountyil.gov)

- Burial Locations: The following page lists the final disposition sites of the indigents buried by the Cook County Medical Examiner's Office.

- Health & Hospitals System – Outpatient Registrations, by Facility, Zip Code, Month – Fiscal Year 2011: Enclosed data represents outpatient registrations including hospital ancillary services

San Francisco, CA [City and County] (data.sf.gov)

- Child Care Subsidies, San Francisco, CA: Data illustrate the total number of state and non-state child care subsidies available as well as the number of children (0-12 years old) that are eligible for subsidies and the difference between these two numbers by zip code in San Francisco.
- HSA 90 Day Emergency Housing Waitlist: Provides the seniority list for entry into HSA 90 day emergency shelter waitlist. The list will be generated on 2/24/14 and updated twice daily.

State Government Data

States administer Medicare, Medicaid and often provide funding for local health and human service programs. Below is a highlight of some state data sets.

Illinois (data.illinois.gov)

- Affordable Care Act (ACA) Enrollment Summary Data: Affordable Care Act (ACA) enrollment data by age, race, gender, and county. (PDF)
- Reportable Communicable Disease Cases, 2010 – 2012: Data provided by the Communicable Disease Section of the Office of Health Protection's Division of Infectious Diseases

New York (data.ny.gov)

- Assisted Outpatient Treatment) Court Orders: This dataset contains the number recipients with AOT petitions and court orders and their length of time on court order, by county, region and statewide.

- Genealogical Research Death Index Beginning 1957: The Genealogical Research Death Index assists individuals with locating New York State (NYS) death records that fall within defined genealogy years, exclusive of New York City recorded death records

City Government Data

Cities are also releasing data on health and social services. Here's some highlights from different cities.

Chicago (data.cityofchicago.org)

- Food Inspections: This information is derived from inspections of restaurants and other food establishments in Chicago from January 1, 2010 to the present.
- Neighborhood Health Clinics (Historical): Former neighborhood health clinic locations, hours of operation and contact information. These clinics were closed or transferred to private management in July 2012
- Infant Death Mortality in Chicago: This dataset contains the annual number of infant deaths annually, cumulative number of infant deaths, and average annual infant mortality rate with corresponding 95% confidence intervals, by Chicago community area, for the years 2005 – 2009

Boston (https://data.cityofboston.gov)

- Asbestos Removal Permits: Boston Public Health Commission Asbestos permitting removal of asbestos.

New York City (https://nycopendata.socrata.com)

- New York City Leading Causes of Death: The leading causes of death by sex and ethnicity in New York City in since 2007
- New York City Health and Hospitals Corporation (HHC) Patient Satisfaction Survey: Patient satisfaction at HHC hospitals is measured by a standardized survey known as the Hospital Consumer Assessment of Healthcare Providers & Systems (HCAHPS). The survey has been validated by the federal Centers for Medicare and Medicaid Services (CMS) as a standard assessment tool for all hospitals throughout the nation.
- Most Popular Baby Names by Sex and Mother's Ethnic Group, New York City: The most popular baby names by sex and mother's ethnicity in New York City.
- Potential Partners

The best civic apps are built through partnerships between technologists, residents, and the people who work on the front lines. Here's a list of potential partners you can work with in your own cities to help build projects that can make an impact.

Health Data Consortium: The Health Data Consortium is a public-private partnership working to foster the availability and innovative use of open health data to improve health and healthcare. This organization is particularly useful for government agencies looking for help opening up health data.

Code for America Health Focus Team: The health focus area works to improve the health of people and their communities. Code for America works with the wide variety of teams that contribute to these outcomes—including city health departments, public health agencies, state offices, and non-profit organizations.

Smart Chicago Collaborative: Smart Chicago's multiple health initiatives provide equipment, training, and information that allow residents to take action to improve their own health. We are strong advocates for promoting open data practices in the healthcare field. Smart Chicago is always happy to talk and share our work.

Local Health Departments: Local health departments are in the trenches on a daily basis working to make their communities healthier and can make great partners. The Chicago Department of Public Health was one of the first city agencies to jump into civic hacking with the Chicago Flu Shot app.

Examples of Health Related Projects

mRelief

mRelief is a site that simplifies the social service qualifying process with an easy-to-use form that can be accessed online and through SMS. Residents can check to see if they're eligible for a variety of programs including food stamps, medicaid, WIC, and more.

EBT Near Me

EBTNearMe is the easiest way to find stores and surcharge-free ATMs where you can use your EBT card in California. It was build by the Code for America Health Team because California welfare recipients pay nearly $20 million per year in ATM surcharge fees partially because there isn't an easy way to find the free ones.

It's an open source project built with public retailer data from the USDA and ATM data graciously shared by the CA Office of Systems Integration.

Foodborne Chicago

Foodborne Chicago uses computers & code to search Twitter for tweets related to food poisoning in Chicago. The system does as much as it can to automatically zero-in on the tweets Foodborne thinks are really about a possible food poisoning case and really coming from Chicago. Then real humans from the Chicago

Department of Public Health review the tweets and @reply back to people with a link back to this page where Foodborne asks for additional information. When they fill out the online form, it becomes a 311 service request to inspect the suspect restaurant.

Chicago Health Atlas

The Chicago Health Atlas a place where you can view citywide information about health trends and take action near you to improve your own health. The site displays large amounts of data from sources like the City of Chicago, State of Illinois, and local hospitals so you can get big-picture views of health statistics in Chicago like hospital admissions, uninsurance rates, cause of death, birth rates. and drill down deep into neighborhoods to see specific information and how it compares to the city overall.

People to follow on Twitter

- @lippytalk: Jake Solomon is a member of Code for America's health focus team and spent time on SNAP benefits so he could better understand the challenges that users face.
- @reedmonseur: Raed Mansour works on #publichealth tech innovations for @ChiPublicHealth like @FoodBorneChi, BU #HealthComm Faculty, APHA Member, @PurdueAlumni & @BUalumni.
- @PublicHealth: Official account of the American Public Health Association: For science. For action. For health.
- @CDCgov: CDC's official Twitter source for daily credible health & safety updates for Centers for Disease Control & Prevention

Safety and Justice Primer

Safety and Justice is a relatively new broad term that used to describe the subject areas of criminal justice, public safety and reform efforts. The term is also used to try to bridge the tension between the need to keep our communities safe while ensuring that the mechanisms used to do that are equitable and fair.

While the topic may be broad, there are some very specific challenges that are on the forefront.

Overcrowding at Jails

When people get charged with a crime, they are sent to jail. If they can secure a bond (a payment made to the court to secure their appearance at trial), then they are released on bail. If they cannot, then they are placed in jail to await their trial at a cost to the government that's often greater than the amount of their bond. (By contrast, Prisons are where people are sent after they have been convicted of a crime – they are not the same thing.)

According to the Vera Institute for Justice, even as crime has gone down the population of jails has increased nationally as economically disadvantaged people fill the prisons because they are unable to make bail. According to the study, 75% of people currently in jail awaiting trial are there for nonviolent crimes like property, traffic, or drug crimes. The overcrowding of prisons largely impacts people in low income neighborhoods and communities of color. Per the NAACP, African Americans are incarcerated at nearly six times the rate of whites.

The other element is that some jails have become dumping grounds for the mentally ill.

Tom Dart is the sheriff of Cook County, where he oversees what he calls the largest mental health facility in the State of Illinois. In

45 states, the largest mental health provider is a prison or a jail.

There are a number of things that governments like Cook County are trying to do to address the issue including electronic monitoring and alternative sentencing (often where the inmate is diverted to a treatment facility instead of incarceration.) However, impending budget cuts may stem progress.

This trend of more people in the system is also occurring at the same time overall crime is trending down nationally.

Juvenile Justice

Another big challenge that people in this area tend to focus on is juvenile justice. Convictions for criminal convictions tend to skew young – particularly in economically depressed areas. The problem is often exacerbated by cuts in school funding and youth programs designed to keep youth away from gangs. In addition, the victims of gun violence are also disproportionately young and male.

Once a person turns 18, there are ways to have a juvenile record expunged. This doesn't just mean the case is removed from a criminal record – expungement means that in the eyes of the law the incident never existed. So, for example, if a person is asked on a job application if they've ever been arrested for a crime they can legally answer "No" if that incident was expunged.

However, in many states this isn't automatic. The resident has to go through a legal process that can be complicated. Even if the resident wasn't found guilty or had their case dismissed, they may still have a criminal record which can prevent them from getting a job, joining the military, or getting into college.

Even when an expungement is successful and in the eyes of the law didn't happen, the Internet has a long memory. In some cases, news stories about an arrest still exist and come up first in Google searches even if the person wasn't convicted. In worst cases, nefarious businesses make money by scraping mugshot pictures from government websites then charging people to remove them. Google has recently changed the search algorithm to push mugshot sites down the results list and the FBI has asked victims of these schemes to file a complaint.

A complex system

Tackling problems in the criminal justice system can seem daunting simply because of the complexity of the system. The criminal justice systems is made up of many government entities at the local, state, and federal level that each handle a different aspect of the system. From a data perspective, each agency has it's own data system and very little of it is handed off from one part to the other. (So, you couldn't take one incident report number and trace it all the way through to somebody being sentenced to prison.)

Data Sources

We've highlighted some data sets that pertain to the area of safety and justice below. However, because the criminal justice system is complex it may be of use to check out Crime and Punishment in Chicago to get a better sense of the different aspects of data available. You can also check out the Sunlight Foundation's excellent coverage on the topic – including a Google Doc that showcases a list of known data resources nationwide.

- National Crime Victimization Survey (US Department of Justice): This research helps the justice department measure how many people are victims of certain kinds of crimes, or the victimization rates.
- Dallas Police Department Call of Service: This data release stands out because it identifies the units responding to each event, a link to map that will display the block from where the call originated, and the status of the call, among other included data fields. However, the data can't be downloaded.

- Chicago Crimes 2001 – Present: This dataset reflects reported incidents of crime (with the exception of murders where data exists for each victim) that occurred in the City of Chicago from 2001 to present, minus the most recent seven days.
- Henrico County (VA) Arrest Reports: Arrest data is searchable, but not downloadable in bulk.
- Cuyahoga County Office of the Prosecutor (Ohio): Includes a performance dashboard that includes dispositions statistics, juvenile dispositions, cold case rapes, heroin deaths and prosecutions, incarceration statistics, and charging statistics.
- King County (WA) Adult Bookings: The Jail Bookings Open Data dataset is provided by the King County Department of Adult and Juvenile Detention. The dataset contains adult jail bookings which have occurred within the last twelve months; specific dates are indicated on the web site.
- Maryland State Courts: Website lists what occurs in every court hearing while also detailing information regarding the officer that arrested the defendant, the prosecutor(s) prosecuting the case, and the judge presiding over the case. (No bulk download)
- Oregon Prison Population: Dates and the count of inmates in the State of Oregon's prison system for the date. Based on a DOC report for 1980 -1993. Based on query of DOC data for 1994-current

Potential Partners

- **Legal Aid:** Legal aid foundations and societies offer free legal services to clients that may not be able to afford these services on their own. Many of the apps that focus on expungement, have legal aid foundations as partners.
- **Code for America Safety and Justice Focus Team:** Code for America partners with local governments to build and grow digital safety and justice services focused on increasing public safety and reducing costs, helping governments transform themselves along the way.

- **Neighborhood Organizations:** Neighborhood organizations play a vital role in building relationships between residents, public officials, and other non-profit organizations.

Examples of Safety and Justice Related Projects

- **Expunge.io:** Expunge.io is a website designed for people with juvenile records in Illinois to kick off the process of expunging, or erasing, those records. It is a youth-led project run by Smart Chicago and controlled by the Mikva Challenge Juvenile Justice Council.
- **Jail Population Management Dashboard:** Jail Population Management Dashboard helps judges and other stakeholders understand the conditions in the metro jail and use this data to visualize how their decisions affect program, facility and inmate outcomes.
- **Convicted in Cook:** Convicted in Cook is a project that analyzes five years worth of conviction data received through the Office of the Chief Judge of the Circuit Court of Cook County by the Chicago Justice Project.
- **Courtbot:** Courtbot is a simple web service for handling court case data. It offers a basic HTTP endpoint for integration with web-sites, and a set of advanced twilio workflows to handle text-based lookup. It's been deployed by Atlanta to help people pay their traffic tickets online instead of waiting in line at traffic court.

People to follow on Twitter:

- @TomDart: Sheriff of Cook County. Bringing an innovative approach to law enforcement in the second most populous county in the US.
- @MarshalProj: The Marshall Project is a nonprofit, nonpartisan newsroom covering America's criminal justice system. Tweets usually from @amandablair.

- @PublicSafety: The Coalition for Public Safety is working to reform our criminal justice system to make it more just, more fair, and more effective.

Transportation Primer

Transportation policy impacts many different aspects of city life. Transportation policy impacts economic development, the environment, and overall quality of life. Transportation is also something that cities, states, and the federal government spend a lot of money on. In 2009, total transportation expenditure by State and local governments reached the record level of $270 billion. Despite these expenditures, the American Society of Civil Engineers gives the country's infrastructure a grade of D+. Specifically, the ASCE is calling for additional investments to repair roads, bridges, and rail infrastructure.

Data Resources:

The US Department of Transportation has over 1600 transportation related data sets at data.gov along with several developer resources. These developer resources include:

- Bureau of Transportation Statistics (BTS) Open Source Data API
- Federal Motor Carrier Safety Administration (FMCSA) SaferBus API
- Federal Railroad Administration (FRA) Safety Data APIs
- Federal Aviation Administration (FAA) APIs
- National Highway Traffic Safety Administration (NHTSA) API
- At the local level, Chicago also has several sets of data related to transportation including:
- Average Daily Traffic Counts
- Daily Boarding Totals for the Chicago Transit Authority
- Historical Congestion Estimates by Region API

- The State of Illinois' Department of Transportation also publishes data on transportation including:
- Average Annual Daily Traffic 2010
- IDOT Rest Areas Map

Potential Partners

The best civic apps are built through partnerships between technologists, residents, and the people who work on the front lines. Here's a list of potential partners you can work with in your own cities to help build projects that can make an impact.

- **City Transportation and Transit Agencies:** Chicago Department of Transportation, New York's MTA, or Boston's MTBA.
- **Regional Planning Authorities:** Regional planning authorities are responsible for helping to plan major transportation projects across wide regions. This is often a requirement for these projects to receive federal transportation dollars. Because of that, these agencies often have data useful for transportation projects or experts who can help with projects. Some examples of these agencies are the Chicago Metropolitan Planning Council, Association of Bay Area Governments, and the Miami Valley Regional Planning Commission.
- **Cycling Groups:** Cycling groups try and advocate for policy reforms to promote cycling in cities. These groups may be familiar with current city policies, popular bike routes, and trouble areas. These groups include Critical Mass, San Francisco Bicycle Coalition, and Bike Austin.
- **Bikeshare Systems:** Many cities are now installing bike share systems to increase cycling in their cities. Many of these bike share systems regularly release data about ridership and usage that can be used to power apps.

Examples of Transportation Related Apps:

- **SpotHero**: SpotHero is a civic startup that efficiently connects parking demand and parking supply. SpotHero allows parking spot owners to earn income by renting their spot when it's unoccupied. SpotHero was the winner of the Apps for Metro Chicago Challenge in 2012. Since then, they've graduated from TechStars/Excelerate , raised 2.5 million in VC funding, expanded to seven cities and has hired twenty people.
- **Chicago Bike Crash Map**: Crash data for Chicago in 2005-2012 where a bicyclist or pedestrian was the first point of impact by a driver's automobile, as collected by responding law enforcement and maintained by the Illinois Department of Transportation.
- **TextMyBus**: Built by Code for America for the Detroit Department of Transportation. This app lets riders text a number to see when the next bus is.

Other resources on the web

Streetsblog: Streetsblog is a national network of blogs that cover transportation issues on a daily basis.

People to follow on Twitter

- @SteveVance – Transportation reporter for Streetsblog Chicago and one of the co-organizers of the Transportation of OpenGov Night Night.
- @Gabe_Klien – Former transportation chief for DC and Chicago.
- @Anthony_Foxx – Current US Secretary of Transportation
- @JoesphKopser – CEO and Founder of Ridescout, a transit app that aggregates data from multiple transit agencies

Part Three

Wins

Here's a series of blog posts I wrote to talk about "wins" in Chicago civic tech. The dates of these posts will vary, but you'll notice a pattern. Our civic tech wins are marked by a collaborative effort that focuses on people.

Smart Chicago and the City Tech Plan

The City of Chicago is a founding partner of the Smart Chicago Collaborative and they have a number of responsibilities in the Chicago Tech Plan. Here's an update I posted regarding the city's most recent technology plan.

June 25th, 2015

At Techweek, City of Chicago Chief Information Officer Brenna Berman announced an 18-month update to Chicago's Tech Plan.

Chicago's first Tech Plan was first launched in 2013 and laid out a strategy to establish Chicago as a national and global center of technological innovation.

Since it's launch, Chicago's civic technology community has made significant progress towards the goals of the tech plan.

As a civic organization devoted to improving lives in Chicago through technology, Smart Chicago is proud to be heavily involved in the implementation of Chicago's Tech Plan.

Here are some highlights from the update.

Next Generation Infrastructure

Chicago is working with internal and external partners to improve the speed, availability, and affordability of broadband across the city. The City is preparing to create a Request for Proposal for companies to design, construct, implement, and manage a gigabit-speed broadband network.

In addition to broadband infrastructure, the city is also working to digitally connect its infrastructure. Part of this includes the launch of The Array of Things project, which will place network of interactive, modular sensor boxes around Chicago collecting real-time data on the city's environment, infrastructure, and activity for research and public use. (You can listen to their presentation at Chi Hack Night here.) You can already get up to the hour updates on beach conditions thanks to sensors maintained by the Chicago Park District. The Department of Innovation and Technology has loaded the information onto their data portal.

Make Every Community a Smart Community

One of the major efforts of the civic technology community in Chicago is closing the digital divide in every neighborhood.

Much of the work in the coming months will focus on Connect Chicago. This citywide effort, led by Smart Chicago in partnership with LISC Chicago, Chicago Public Library, World Business Chicago, and the City of Chicago's Department of Innovation and Technology aligns citywide efforts to make Chicago the most skilled, most connected, most dynamic digital city in America.

Here's more from the Tech Plan about the program:

As part of this initiative, program partners are creating a profile of a fully connected digital community that can be used as a benchmark and will provide best-practice toolkits and other resources to help all Chicago communities reach this benchmark.

Another big part of the City's strategy to close the digital divide in Chicago involves the Chicago Public Library. Libraries around the city already function as public computing centers and now they provide Internet to Go – a program where residents can check out laptops and 4G modems so that they can access the internet at home.

The City of Chicago and the civic tech community is also heavily focused not only access, but on digital skills. The Chicago Public Library's Cybernavigator Program will be expanded and Chicago Public School is working on implementing computer science curriculum at all schools.

On our end, Smart Chicago is working with Get In Chicago to run a youth-led tech program this summer. The conceptual model for this program is "youth-led tech", which means teaching technology in the context of the needs & priorities of young people. Youth will learn how to use free and inexpensive Web tools to make websites and use social media to build skills, generate revenue, and get jobs in the growing technology industry. They will also learn about all sorts of other jobs in tech— strategy, project management, design, and so on.

Effective Government

The City of Chicago's Department of Innovation and Technology is also making great progress in using data to help city government be more efficient and effective. One of their first projects, WindyGrid, is a geospatial Web application designed by the City's Department of Innovation and Technology that strategically consolidates Chicago's big data into one easily accessible location. WindyGrid presents a unified view of City operations—past and present—across a map of Chicago, giving City personnel access to the city's spatial data, historically and in real time, to better coordinate resources and respond to incidents.

The City of Chicago will be open sourcing the project later this year on their Github page.

That's not the only open source project that the city has on the books. Chief Data Officer Tom Schenk Jr. recently spoke at Chi Hack Night to talk about their new system to predict the riskiest restaurants in order to prioritize food inspections. The system has found a way to find critical food safety violations seven days faster. Aside from the important aspect of less people getting sick from foodborne illness in the City of Chicago, there is another very important aspect of this work that has national impact. The entire project is open source and reproducible from end to end.

Since the release of the Tech Plan, Smart Chicago has been working with the Chicago Department of Public Health on the Foodborne Chicago project. Foodborne listens to Twitter for tweets about food poisoning and converts them into city service requests. The Tech Plan update has some results from the project.

A study of the system, published by the Centers for Disease Control, found that during March 2013 – January 2014, FoodBorne Chicago identified 2,241 "food poisoning" tweets originating from Chicago and neighboring suburbs. The complaints identified 179 Chicago restaurant locations; at 133 (74.3%) locations, CDPH inspectors conducted unannounced health inspections. A total of 21 (15.8%) of the 133 restaurants reported through FoodBorne Chicago failed inspection and were closed; an additional 33 restaurants (24.8%) passed with conditions, indicating that serious or critical violations were identified and corrected during inspection or within a specified timeframe.

Chicago's open data portal is also getting expanded as part of the updated Tech Plan having grown by more than 200 data sets over the last two years. Chicago was the first City to accept edits to select data sets through the City's GitHub account.

Open311 is also getting an upgrade with the city undergoing a procurement process to build a new 311 system. As part of the process for upgrading 311, the new system will go through user testing through the Civic User Testing Group.

Civic Innovation

A big part of the city's strategy around civic innovation is supporting the work of civic technologists here in Chicago. As part of the Tech Plan, Smart Chicago will continue to provide resources to civic technologists like developer resources, user testing, and financial support to civic technology projects.

The Tech Plan also calls out our work with the Chicago School of Data. The two day experience was wholly based on the feedback we received from dozens of surveys, months of interviews, and a huge amount of research into the work being done with data in the service of people. If you missed the conference, here are some of the key takeaways.

The Civic User Testing Group also plays a part in the Tech Plan and has recently been expanded to include all of Cook County.

Chicago Chief Information Officer Brenna Berman stated that Chicago has the strongest civic innovation community in the country. A large part of that community has been the Chi Hack Night, now in it's fourth year with attendance now reaching over 100 people regularly.

Technology Sector Growth

One of the thorniest issues for civic technologist is the issue of government procurement. One of the things that the city has been doing is meeting with different groups to talk about ways the city can make it easier to buy products and services from smaller business and startups.

As part of the Tech Plan, the City of Chicago is taking this on directly. Here's the quote from the Tech Plan:

This summer, DoIT will release a Request for Qualifications for start-up and small-sized companies to join a new pool of pre-qualified vendors eligible for future City procurement opportunities. Companies who are deemed qualified will be placed into a pool and receive access to City contract opportunities in the areas of software application development and data analytics.

To further decrease the barriers facing smaller-sized companies in competing for City business, the City has modernized its insurance requirements to allow for pooled insurance plans. Start-ups that are members of an incubator, such as 1871, or smaller companies that come together for a group insurance plan, may now meet the City's insurance requirements as a group. Insurance requirements were identified as a barrier to conducting business with the City in a series of listening sessions conducted over the past year with these companies.

This is a huge opportunity not only for civic tech companies, but it will enable the city to take advantage of the innovation coming out of these companies.

Side effect of civic tech partnerships may be healthier communities

---

August 7th, 2013

One of the success stories in Chicago's civic innovation community is the rapid spread of health related apps that have come out of both the volunteer civic technology community and paid development efforts. This started last year with Tom Kompare's Chicago flu shot app that helped Chicago residents find free flu shots near them. (Later on, this flu shot app spread to Boston and Philadelphia.)

Professionally, Kompare is a web developer with the University of Chicago. In his spare time, he's one of the most active civic technologists in Chicago.

Kompare's flu shot app was just the start of the Chicago Department of Public Health partnering up with civic technologists on a number of projects including Foodborne Chicago, the Chicago Health Atlas, and Tom Kompare's newest app Back to School.

Back to School is an app built for parents to make sure that their child has the immunizations they need to go back to school. CDPH hosts several immunization events for school children throughout the city and the apps helps parents find events near them. This will be also good trial run for the larger immunization effort that CDPH will run this fall.

Not only do these two apps use the same data format, this data format is now a proposed national standard. Shortly after the redeployment of the flu shot app in Philadelphia earlier this year, Philadelphia Chief Data Officer Mark Headd began an effort to develop a national standard for flu shot data. This effort included input from both government officials and civic technologists from Chicago, San Francisco, Austin, Oakland, and other cities. By helping to set up one standard, both of these open source apps are deployable in any city that elects to conform to the standard.

This is not the first app that the Chicago Department of Public Health has partnered with. CDPH also partnered with the Smart Chicago Collaborative to run the Foodborne Chicago app. The app, which was made possible through a variety of efforts, listens to Twitter for reports of food poisoning and then prompts the author to a web page that reports food poisoning to 311. Once it's reported the 311, the city can then dispatch a health inspector to that location.

The Chicago Health Atlas, an app that shows health trends and local resources in Chicago, was another app that resulted from community partnerships. Initially, the Atlas was built on an existing partnership by informatics researchers at five major academic health centers in Chicago: University of Illinois at Chicago, Stroger Hospital, University of Chicago, Northwestern University, and Rush University. Since that time, the Chicago Health Atlas has been expanded to include data and researchers from the Chicago Department of Public Health. The site itself was built by Chicago civic app firm Datamade and also uses data from civic startup Purple Binder.

And it's not just formal partnerships that are producing health related apps. As part of a summer internship program, Chicago Spanish newspaper Vive Lo Hoy hired its first web developer Wilberto Morales. Morales worked with food inspection data provided by the City of Chicago to built eatsafe.co.

Eatsafe.co helps residents find out how the restaurants near them fared during their last food inspection. This open source app includes information on how food inspections in Chicago work. (Unlike some other cities, Chicago's system is pass-fail and not by letter grade.) This is Hoy's first app and they plan to continue building more apps to help address community issues.

Morales learned to code, not from a formal computer science program, but by being a part of FreeGeek Chicago's Supreme Chi-Town Coding Crew. FreeGeek Chicago, a Humboldt Park non-profit organization, helps the community by recycling and repairing old electronics. FreeGeek trains volunteers on computer repair techniques and offers the opportunity for residents to earn a refurbished computer through their Earn-A-Box Program. In March, FreeGeek' Chicago's members decided to launch a program to teach residents how to develop web applications. It's certainly been a big success, as evidenced by the Supreme Chi-Town Coding Crew winning the Chicago Migrahack with their app Finding Care.

So why is Chicago producing so many high quality health apps so quickly?

Lesson 1: Open data is fuel

Data is fuel. None of these applications would be possible without the City of Chicago developing an open data policy and executing on it. Chicago has more data sets than any other city. Their deep involvement with the civic technology community allows the city's data team to meet the needs of civic innovators creating apps that serve the community.

Lesson 2: Partnerships are extremely important

The Chicago Chi Hack nights attract some of the city's best geeks to work at the intersection of technology and civic problems. However, if the hack nights and other civic innovation efforts only attract web developers, designers, and data gurus we end up missing a vital piece of the puzzle: the neighbor that we're trying to help.

By partnering with city departments, non-profit organizations, and community organizers – the community is able to develop apps centered on the civic problem in a way that helps the people who are working in the trenches. Both volunteer efforts like

OpenGov Chicago and the Smart Chicago Collaborative are continuing to do outreach with civic organizations to help foster partnerships between technologists and community activists.

Lesson 3: Grow your own talent

The civic innovation community is growing – but not nearly fast enough. To meet the challenges brought on by the Great Recession, we need more people with technology skills necessary to grow the civic technology space. Efforts by FreeGeek Chicago and the Englewood Codes project by Teamwork Englewood are helping to create more technologists in the city's neighborhoods.

Lesson 4: Solve real problems

Lastly, it's important that the civic innovation community solve real problems that matter to real people. Part of the side effect of having strong partnerships in both government and the neighborhoods is the learning what is happening on the front lines of civic work. The Back to School app came about as a result of Tom overhearing a conversation about CDPH's back to school campaign while working on the update for the flu shot app. Having civic technologists listening to not only tech issues, but issues like education results in web app like schoolcuts.org being created.

Building civic apps is more than just the code – the real secret to success is community.

Englewood Codes

August 29th, 2013

Tonight, 25 teens from Englewood Codes will demonstrate their websites at Kennedy King College. Englewood Codes is a 10 week summer program run by Demond Drummer of Teamwork Englewood. The program teaches kids not just web development, but teamwork and leadership.

Teamwork Englewood was first formed in 2003 as part of the New Communities Program sponsored by LISC – Chicago and the MacArthur Foundation. They've done work in the Englewood neighborhood to help increase the digital adoption rate. This includes helping to run the Englewood Portal and helping to connect local residents to technology.

This included an effort in 2011 to teach 15 teens how to code over the course of six weeks.

The effort generated a lot of buzz and youth in the neighborhood started asking about the program again in 2012. In 2013, an effort was launched on Kickstarter to run the program again – this time for ten weeks and using Raspberry Pi's.

Getting the project launched:

The project had a great amount of support throughout the city and the community including the Department of Innovation and Technology, After School Matters, LISC-Chicago and the City Colleges of Chicago which donated space for the program.

World Business Chicago had recently started a curated Kickstarter page to highlight projects that involved community development called Seed Chicago. The project was featured on the page and began to buzz on social media. They reached their fundraising goal with plenty of time to spare.

But Teamwork Englewood didn't stop there – they created a stretch goal to expand the number of students to 30 – and received the funding to it. In total, the project had 167 backers. (Including $1,000 from Smart Chicago as Demo Day host.)

What Englewood Codes Accomplished

During the 10 week course, the students learned HTML, CSS, Bootstrap and working with GitHub. However, the program didn't just teach code the program also taught teamwork and leadership. Listen below as Demond describes how they formed the students into teams and encouraged the group to work together to solve problems.

The programs also made great use of the GitHub platform. Because GitHub can be used using just brower tools, it didn't require Englewood Codes to install anything on the computers. Students created different branches of their work, forked other student's codes, and used the gist feature to post notes and staff used it to post lesson plans.

Lessons Learned

Demond also shared some lessons learned from running the program.

- Small is beautiful: Small class sizes allowed for more one on one interaction.
- Go deep VS going broad: Going more in-depth on topics and not rushing the course.
- Debug bad learning habits: There's always something to do and learn, you can go ahead of the teachers, you can learn from your peers!
- Teamwork makes the dream work!
- Make it real: Connect what we're learning to the real world.

- Live in the cloud: GitHub was amazing because Demond could see every student's code.
- Dedicated space is ideal

What's next for Englewood Codes

Englewood Codes will continue as a fall after school program. Englewood Codes will also help jumpstart a middle school computer club. The program will also connect students with professional mentors.

Englewood Codes will also take on a greater advocacy role. In order for Chicago to grow their own talent, Chicago will need to maintain computer science teachers in neighborhood schools. Englewood Codes will also push for a federal mandate for computer science classes in schools.

City of Chicago launches OpenETL Toolkit

The City of Chicago's Department of Innovation and Technology is responsible for handling all of the city's data. In addition to producing a vast about of open data on their data portal (data.cityofchicago.org), they also are harnessing the power of their own data to build a predictive analytics platform. As part of this effort, the city has released a number of their tools on their GitHub account (github.com/Chicago) including the OpenETL toolkit. This was one of the most popular posts I ever wrote for Smart Chicago.

September 24th, 2014

The City of Chicago's Department of Innovation and Technology released the OpenData ETL Utility Kit at the Code for America Summit this morning. The ETL Utility Toolkit will give cities the same tools Chicago uses to get data from their own internal systems and unto their data portals.

We'll explain why this is such a big deal and what ETL is below!

What's an ETL?

ETL stands for Extract, Transform and Load. It's shorthand for a system that takes data from one system and puts it someplace else. Like fancy plumbing for data.

So, say that the city tows your car. At some point, the city enters your vehicle's information into a city system. On the DoIT's end, they use an ETL system to take the data from that system, transforms it into a format that they can then put here – on the city's data portal. Once it's on the data portal, civic technologists can work with it.

To do this, the City of Chicago has deployed a number of ETL tools. They've now open sourced those tools as part of a toolkit that any city will be able to adopt for free.

Features

Here's how the team describes the app on GitHub:

ETL Utilities for an Open Data Program

This toolkit provides several utilities and framework to help governments deploy automated ETLs using the open-source Pentaho data integration (Kettle) software.

Namely, this toolkit will assist with:

- Load data from a database an load it to a Socrata data portal
- Steps to integrate with an email server to provide e-mail alerts on the outcome of ETL scripts
- Handles deployment issues when using multiple operating systems during development
- Utilities to allow administrators to quickly analyze the log files of ETLs for quick diagnostics
- The ETL framework is organized so each function can be modified in one file that is used by all ETLs. This provides for easier maintenance, upgrading, and modification over hundreds of ETLs.

Features

- Open source at the core – this framework can be deployed using Kettle, an open-source ETL software. Pentaho also provides telephone support and training if desired.
- Compatible with multiple data sources – this ETL framework can be used with a variety of data sources, including a range of databases (MySQL, PostgreSQL, Oracle, SQL Server, and variety

of NoSQL), APIs, text files, etc.

- Compatible workflow for multiple operating systems – ETLs can be developed and deployed across multiple operating systems. ETLs can be developed on a Windows environment and deployed on Linux
- Helpful utilities – includes several scripts to help users quickly analyze log files

Four Days After Divvy bike share launch an app prototype and an API presented at hack night

July 8th, 2013

Just four days after the launch of the Divvy Bike Sharing program, Chicago's civic technology community had already put together a prototype app and an API using bike share data.

Chicago Chi Hack Night played host to representatives from both Alta Bicycles (The company that manages Divvy) and the Chicago Department of Transportation to talk about the new bike share program and the data that's available.

The first data set, available at (http://divvybikes.com/stations/json), gives information on real-time usage statistics and availability of bikes per station.

This data set reflects the same format from other bike share systems across the globe. This enabled Chicago's data to be merged into other bike apps such as Oliver O'Brian's Global Bike Share Map.

Locally, Ian Dees and Steve Vance have been hard at work building applications with this new data set. They have put together google document that contains a list of resources for the civic innovation community who are interested in working with bike share data. Dees has built an API that makes it easier to access the bike data. Vance has already taken advantage of this API to provide bike share information in his Chicago Bike Guide. (As a reminder, this was presented to Chi Hack Night just FOUR DAYS after the launch of Divvy.)

Steve Vance and Ian Dees discuss apps that are built using bike share data

Daniel Gohlke from Alta bikes answered questions about the in and outs of the data as well as plans to release information on the most active routes and where people are biking. Gohlke's stated that in most cities this data is released quarterly, but if CDOT wants to release this data at an increased frequency that it should be possible.

Question and answer session with representatives from Divvy and CDOT

Given that Divvy was only launched for a little over 2 weeks ago, we're excited about what's to come in terms of potential uses for this data. John Tolva, Chief Technology Officer for Chicago, would like to see data from different transportation systems integrated into a single app:

"The speed with which our civic innovation community seized on Divvy data gives me great hope that a truly multi-model, hyper-local journey planning service — across cabs, public transit, shared bikes — can be assembled fairly easily. Most of the data is there. All that's needed is an intelligent stitching-together."

For those interested in working on bike share data, you're invited to attend the next Chi Hack Night – every Tuesday at 6:00pm inside Merchandise Mart.

Largelots, a working app for Chicago neighborhoods

April 29th, 2014

On the Chi Hack Night on April 15th, Demond Drummer presented at OpenGov Hack Night about the Large Lots Program. This city program is designed to allow residents to purchase vacant lots in their neighborhoods for just a single dollar in an effort to help revitalize the neighborhood.

The Large Lot program allows the city to transfer the vacant lots in the Green Healthy Neighborhood Program area to nonprofits or residents that own property in the same block as the vacant property. Previously, the program was set up so that you had to be on an adjacent lot. To help increase the number of lots transferred, the city and community groups began working together to update the program.

The program was designed community feedback in mind. Teamwork Englewood and Resident Association of Greater Englewood (R.A.G.E.) spearheaded the community outreach around the program which launched required a tremendous amount of community organizing and planning.

Once the policy had been updated, Teamwork Englewood reached out to Chicago civic tech company DataMade to build the site. Fueled by freshly released Cook County parcel data and extensive knowledge of how to apply to purchase a lot, the site provides all the information needed for residents to purchase large lots.

A user enters in their address to see a map of all available lots near them. If the user is interested, the site will provide detailed instructions on how to purchase the lot with links to the relevant sites. (*Future iterations of the app would allow users to fill out the applications all from inside the application.)* Once the pilot is complete, the site then will show all of the lots that were purchased under the program.

Press

The Largelots program was featured in a number of press publications including:

- WBEZ's Afternoon Shift
- The WBEZ Public Data Blog
- DNAinfo
- Chicago Sun-Times
- Curbed Chicago

Results of Largelots.org

This round of the Large Lots program was a pilot. In the short time the application period was open the City received over 400 applications to purchase lots. The site itself received over 6,300 pageviews.

More importantly, a great many number of vacant lots will now be purchased and used by people in the neighborhoods. Vacant and abandoned buildings remain a thorny problem for the city and initiatives like this are a great step towards solving them.

Additionally, apps like these show that there is a marketplace for civic technology working in tandem with city governments and the non-profit sector.

mRelief – Mobilizing social service relief in Chicago

November 11th, 2014

Last week marked the launch of mRelief, a site that simplifies the social service qualifying process with an easy-to-use form that can be accessed online and through SMS. Residents can check to see if they're eligible for a variety of programs including food stamps, Medicaid, WIC, and more. Here's the press release by mRelief creators on day of launch.

mRelief is made by an all-woman team hailing from different backgrounds and walks of life dedicated to making an impact with technology. Smart Chicago has supported mRelief under our CivicWorks Project funded by the John S. and James L. Knight Foundation.

mRelief

mRelief is already deployed at the Martin Luther King Jr Community Services Center. The City of Chicago's Department of Family and Support Services run the Community Service Centers. DFSS Community Service Centers help individuals and families in need access a wide range of resources from shelter, food and clothing to domestic violence assistance, job training/placement and services for the formerly incarcerated. Staff members are using mRelief to help streamline the process of evaluating their eligibility.

The mRelief team has also partnered with Purple Binder to refer residents to other useful local resources if they are ineligible for public assistance.

How mRelief Works

mRelief works by having users fill out a quick form to see if the user is eligible. If the users eligible, they're then connected to that office. If the user isn't eligible, the system will try to find alternative programs using Purple Binder.

mRelief will continue to work with Smart Chicago post launch to conduct user testing and enable SMS messaging for the site so that those without access to a computer can take advantage of the site. (*This was done soon after the post was written and now users in Chicago can access the mRelief site through just SMS.*)

About the team

mRelief is an all-woman startup out of Chicago launched this past summer. The founding team members are Rose Afriyie, Genevieve Nielsen, and Marina Goldshteyn.

The team is alumni of the Starter League a successful coding school in Chicago. Project Manager Rose Afriyie is a former Googler and White House intern and holds an MPP from the University of Michigan. Chief Technology Officer Genevieve Nielsen was 2014 valedictorian of Davidson College and graduated with an economics degree. She has successfully launched freefoodfinder.co. Lead designer Marina Goldshteyn, was the lead designer on healthcare website https://www.humana.com/ and also designed http://jerrygarcia.com.

You can find out more about mRelief at mRelief.com.

The Chicago Area Red Cross shows us the power of open source

On Tuesday November 18th, Jim McGowan with the Red Cross of Greater Chicago gave a presentation at Chi Hack Night about their open source project: DCSops.

The Chicago Red Cross uses DCSops to manage their situational awareness information and dispatch volunteers to an incident. This is a huge change from January when they were using carbon paper to record information about incidents.

Jim McGowan first encountered Chicago's civic hacking community after attending the first National Day of Civic Hacking event in 2013. The system they were currently using to manage Volunteer Connection was cumbersome and the organization would have to do information management by spreadsheet. McGowan then started attending Chi Hack Night and began to be inspired by all the work being done.

McGowan began working with a developer in San Diego to create DCSops. This was done as the local chapter of the Red Cross and not the 'big' national organization – primarily because it was easier to ask for forgiveness rather than permission. The bet paid off, with DCSops now expanding to multiple chapters across the country. What's more, the app is designed, coded and deployed using volunteers.

The app is responsively designed so that it can be used in the field on smart phones and tablets. This is important because in order to receive cash assistance the Red Cross has to have a signature. With the app, people can simply sign with their finger instead of the carbon paper. The app also saves information on the tablet or smartphone so that if there's no internet connection it can be uploaded later.

In practice, when an incident occurs the Red Cross dispatcher can create an incident within seconds. Once that occurs, the app pulls from Volunteer Connection to get all of the volunteers that are on duty. The dispatcher can select who they want to recruit and send a text message to all potential volunteers. Volunteers can text back that they've accepted the assignment and then check in once they arrive. This information is then fed back into the DSCops system. Additionally, since Twilio now has MMS messaging volunteers can take pictures of an incident and send that back to headquarters.

The Chicago Red Cross is now looking at ways to help grow and sustain the app. This also includes an idea to create a Red Cross Labs so that the Red Cross can hire full time staff to help with tech projects.

Aldertrack: Your Guide to the 2015 Chicago Elections

February 4th, 2015

At the February 3, 2015 Chi Hack Night, Jimm Dispensa talked about Aldertrack – A project to help residents of Chicago follow the 2015 Chicago Aldermanic and Mayoral elections. The site offers race forms, analysis, webinars, as well as providing extensive social media coverage of each race.

History of Aldertrack

The site was first started by Dispensa in 2007 as a text only website that tried to gather details about Chicago's elections and sort them out by ward.

In 2010 Mike Fourcher became a part of Aldertrack and the group partnered with the (now defunct) Chicago News Cooperative to create the first version of Early & Often for the 2011 Chicago elections. (This may sound familiar since the Sun-Times' had it's own Early & Often site that was purchased from the Chicago News Cooperative.)

For this year's elections, Ramsin Canon has also joined the team and is helping to make the site a more complete guide to this year's elections.

Using tools to make the site cheap to run

The Aldertrack team uses several tools to make the site inexpensive to run. Here are the ones the teams lists as indispensable:

WordPress: WordPress is a popular blogging platform that's simple to use. (We use WordPress as well!)

Mailchimp: Mailchimp is a platform that allows users to send out email campaigns. Aldertrack uses it to send out daily email updates about the election.

Slack: Slack is an internal communication tool that organizations use to communicate with each other in real time. It's similar to the old IRC chat rooms, but it's got a much better user interface and can integrate with other platforms like GitHub.

Stripe: Stripe is a payment system that integrates easily with apps. If you've ever used Lyft, you've used Stripe

Join.Me: Join.me is a screen sharing tool that the Aldertrack teams uses for Webinars.

data.cityofchicago.org: The Aldertrack team also makes extensive use of the City Data Portal

The Aldertrack Team also has several tools that they don't love, but use anyway including

- Microsoft Word and Excel
- Adobe Acrobat for PDF reading
- The Chicago Board of Elections Website: The team uses the site to get information, but says it's not the most user friendly site.

There were also sites that the team called useless such as Facebook and printing their own newsletters. Dispensa said that while people will print things out themselves, it's not cost effective to print for a medium-sized run of things. The team at Aldertrack also finds Facebook useless for small audiences.

Lessons Learned

Dispensa shared several lessons that he's learned in running the Aldertrack. The first few lessons revolve around knowing their audience. Dispensa says the audience for Chicago politics isn't as big as you think and that their more interested in raw data than analysis.

Dispensa also says it takes a lot of effort and inside knowledge of local politics to produce quality content for political junkies.

The team also found that traditional advertising doesn't do much for increasing engagement among political junkies. They also found that exposure on other media hasn't driven up readership or sales either.

The Aldertrack team found that while video doesn't get many eyeballs – it does make the organization seem more professionals. For their webinars, the team uses two iPhones and a lapel mic.

Dispensa also says that more open government data has leads to more, not less, information arbitrage opportunities.

The City of Chicago unveils predictive analytics model to find critical food safety violations faster

May 26th, 2015

City of Chicago Chief Data Officer Tom Schenk Jr spoke at last week's Chi Hack Night to talk about their new system to predict the riskiest restaurants in order to prioritize food inspections – and has found a way to find critical violations seven days faster.

The problem with the way that most cities conduct food inspections is that by law they have to inspect all of them. However, the number of restaurants far outweigh the number of inspectors. In Chicago, there's one inspector for every 470 restaurants. Since they have to inspect them all, the normal way of doing this is random inspections. However, the team knew that the residents wouldn't get foodborne illness at random restaurants – they would get sick from those few restaurants that don't follow all the rules.

The Department of Innovation and Technology partnered with the Chicago Department of Public Health and staff from Allstate Insurance to see if they could use analytics predict which restaurants would have critical violations. (Side note: It's a brilliant move on the part of the City and the Allstate to contribute volunteer hours using something that actuaries specialize in.) Some of the data sets used to make these determinations were:

- Establishments that had previous critical or serious violations
- Three-day average high temperature (Not on the portal)

- Risk level of establishment as determined by CDPH
- Location of establishment
- Nearby garbage and sanitation complaints
- The type of facility being inspected
- Nearby burglaries
- Whether the establishment has a tobacco license or has an incidental alcohol consumption license
- Length of time since last inspection
- The length of time the establishment has been operating

All of the data, with the exception of the weather and the names of the individual health inspector, come directly from the city's data portal. (Which builds on the city's extensive work in opening up all this data in the first place.) When factoring all of these items together, the research team was able to provide a likelihood of critical violations for each establishment, which was developed to prioritize which ones should be inspected first.

In order to test the system, they conducted a double-blind study over a sixty day period to ensure the model was correct.

The system has gotten rave reviews and coverage from a number of publications and entities including Harvard University, Governing Magazine, and WBEZ's Afternoon Shift.

Aside from the important aspect of less people getting sick from foodborne illness in the City of Chicago, there is another very important aspect of this work that has national impact. The entire project is open source and reproducible from end to end. We're not just talking about the code being thrown on GitHub. (Although, it is on the city's GitHub account.) The methodology used to make the calculations is also open source, well documented, and provides a training data set so that other data scientists can try to replicate the results. No other city has released their analytic models before this release. The Department of Innovation and Technology is openly inviting other data scientists to fork their model and attempt to improve upon it.

The City of Chicago accepts pull requests as long as you agree to their contributor license agreement.

Having the project be open source and reproducible from end to end also means that this projects is deployable to other cities that also have their data at the ready. (Which, for cities that aren't, the City's also made their OpenETL toolkit available as well.)

The Department of Innovation and Technology has a history of opening up their work and each piece they've released (from their data dictionary to scripts that download Socrata datasets into R data frames) builds on the other.

In time, we may not only see Chicago using data science to improve their cities – but other cities building off the Chicago model to do so as well.

Does your building not recycle? Report it here!

February 5th, 2015

The Environmental Breakout Group at Chi Hack Night has been working on a site to report buildings that don't recycle. Chicago's ordinances require buildings with five or more rental units to provide recycling. However, the city doesn't always enforce this law leading some landlords to not provide recycling.

Claire Micklin, Ben Wilhelm, and Alex Kahn put together to help residents report buildings that don't recycle.

The team will be using the data to visualize reports of buildings that don't recycle and provide a hub for residents seeking more comprehensive recycling services. Already, the site has received 900 reports on 750 locations in the first week that it launched. (The site was getting hit so much that it was running into Google's API limits – and so they started using Smart Chicago's Google API account.)

Micklin first started the project after seeing so many of her neighbors blue bins filled up constantly because neighboring buildings weren't providing recycling.

Micklin says she found working with the Environment Breakout to be really helpful for finding resources and coming up with ideas on how to do it. Scott Beslow, who helps lead the breakout group, helped put the bones together in terms of constructing the site and getting developers interested. The team used the OpenCity Apps Github repository to collaborate on the app.

The site was announced at last week's Chi Hack Night to rave reviews and has appeared on WBEZ, the Mike Nowack Show, and DNAinfo.

The group hopes to use the information to lobby for stronger enforcement of the recycling ordinance. The team will continue to work on the app during Chi Hack Night that happens every Tuesday at 6:00pm at the offices of Braintree.

Part Four

Lessons Learned

The next few chapters go over some lessons learned from running different projects for the Smart Chicago Collaborative.

Lessons Learned from the CivicWorks Project

The CivicWorks is a program funded by the Knight Foundation and the Chicago Community Trust to spur support for civic innovation in Chicago. The program goals are to produce 200 content pieces, 5 apps that solve government problems, and 5 apps that solve community problems.

The project had five top-level goals:

- Work with government entities to make more public data available
- Create an online publication providing news and analysis on open government
- Help government agencies to use their own data to improve public services
- Engage non-profit organizations on efforts to find community solutions that are based on public data and data collected by their own organizations
- Identify technology solutions benefiting government performance and community solutions that can be sustained by the private markets

Descriptions of Project and their Results

Illinois OpenTech Challenge

The Illinois Open Technology Challenge to bring governments, developers, and communities together in a common mission to use public data and create digital tools that will serve today's civic needs and promote economic development. Smart Chicago worked with government officials to publish 138 new datasets (34 in Champaign, 15 in Rockford, 12 in Belleville, and 77 from the 42 municipalities South Suburban Mayors & Managers Association). Smart Chicago also worked with developers in meetups all over the state-—in 6 locations in 4 cities with 149 people. Smart Chicago also preformed outreach in each of our communities to reach regular residents with needs that can be addressed through data and technology.

SWOP/LocalData

With LocalData, the Southwest Organizing Project's process of inventorying vacant buildings is faster, collaborative, and more complete. When we first engaged with SWOP, there were conducting the surveys of the area on paper and transcripting the information into Microsoft Excel - then used Microsoft Trip Planner to draw up the maps. Now,the survey is done right from the a smartphone and the data is uploaded into a central site. This data, which is more accurate than the city's current dataset of abandoned buildings, is proving to be valuable to efforts such as the Cook County Land Bank project — and to other organizations across Chicago.

Affordable Healthcare Act Outreach App

The Affordable Care Act Outreach App is a text-based tool we created to make it easier for Navigators from our partner LISC-Chicago to help residents sign up for the Affordable Healthcare Act through the healthcare.gov site. During the enrollment period, LISC's Health Care Navigators have collectively enrolled 2,975 Chicagoans. Thanks to the Wufoo platform, LISC Chicago was able to track how their navigators did. During these outreach events or one-on-one appointments, navigators have made meaningful contact with nearly 27,000 residents.

Textizen

Textizen is an SMS based survey tool that civic organizations can use to get resident feedback. Organizations can create a survey campaign and then place the survey options on posters, postcards, or on screens during live events. Organizations can then receive real time feedback as people text in their answers.

We used Textizen to aid the campaigns of Chicago Defender Charities, CDOT, Chicago Department of Cultural Affairs, and the Metropolitan Planning Council. Two of these organizations, CDOT and Metroplanners, ended up purchasing their own Textizen plans. The CDOT campaign was the largest Spanish language response in the Textizen's history and the Chicago Defender Charities had the largest ever response to a poll.

WBEZ Public Data Blog

The WBEZ Public Data Blog is dedicated to examining and promoting civic data in Chicago, Cook County and Illinois. WBEZ is partnering with the Smart Chicago Collaborative to provide news and analysis on open government by producing content items that explain and tell stories hidden in public data.

In total, the WBEZ Public Data Blog created 124 blog posts. More importantly than the blog, was that this relationship ended up putting Chicago civic technologists on WBEZ's afternoon shift. There are over thirty stories on WBEZ's tech shift featuring civic technology - many of them being civic hackers who build projects on at hack night. This allowed people who normally don't get on the radio to talk about their work.

While Afternoon Shift was shut down by WBEZ, we do note that the Public Data Blog is continuing forward.

Chicago Works for You

Chicago Works For You is a citywide dashboard with ward-by-ward views of service delivery in Chicago. The project is our most robust use of the City's Open311 system.

Foodborne Chicago

Foodborne Chicago is a website that connects people who complain about food poisoning on Twitter to the people who can help them out— the Chicago Department of Public Health. Last year, Foodborne Chicago classified over 2,600 tweets related to food poisoning in Chicago, which led to 233 food poisoning reports submitted to CDPH. From those reports an additional 150 restaurants and foodservice inspections occurred.

The app was also a precursor to the city's new system for using predictive analytics to identify new cases of critical food violations seven days in advance.

Crime and Punishment in Chicago

The Crime and Punishment in Chicago project provides an index of data sources regarding the criminal justice system in Chicago. We aggregated sources of data, how this data is generated, how to get it, and what data is unavailable. This project is a key way we are using the Civic Works grant to use data journalism to uncover the value of data and cover the stories behind the data.

The project also supported the work of FreeGeek Chicago and the Chicago Justice Project.

The site also helped to inspire the Sunlight Foundation's Criminal Justice Project that uses Crime and Punishment as a model to catalogue criminal justice data from all over the nation.

LocalWiki

LocalWiki is a hyperlocal wikipedia designed to let anyone edit and update information about their city. LocalWiki Chicago is a part of our CivicWorks project, funded by the Chicago Community Trust and the Knight Foundation. Smart Chicago paid for the seeing of content in LocalWiki combined with the Chicago Public Library.

Convicted in Cook

As a follow-up to the Crime and Punishment in Chicago project, Smart Chicago partnered again with the Chicago Justice Project and FreeGeek Chicago to clean up and analyzes five years worth of conviction data received through the Office of the Chief Judge of the Circuit Court of Cook County by the Chicago Justice Project.

The project also help to support the work of the Chicago Justice Project, whose data ended up fueling follow up work like Chicago's Million Dollar Blocks.

Promptly.io / A Safe Haven

As part of the CivicWorks Project, we've helped launch a partnership between Promptly.io and A Safe Haven to provide the Promptly text-based followup services to A Safe Haven clients. One of the challenges that face A Safe Haven is following up with their clients after they've left the program. Clients often move and it's not always easy to maintain contact people once they've left. If A Safe Haven was aware that somebody needs additional services - such as career counseling - then they can offer it if It's also more difficult to track progress over the long term.

Through the CivicWorks Project, we're provided A Safe Haven a license for Prompt.ly. A Safe Haven will use the software to text follow-up messages to their clients to see how they're doing. This will enable A Safe Haven to both track the progress of their former clients as well as reach back out to residents who may need further assistance.

mRelief

mRelief is an app that helps Chicagoans determine their eligibility for government benefits.

Smart Chicago provide $15,000 through our CivicWorks Project, which is funded by the Knight Foundation. We also host their text-messaging platform on our Twilio account under our Developer Resources program. We also conducted a CUTGroup test on the product in early 2015.

mRelief has grown tremendously and has added services, locations, and the ability to text to their app. Additionally, they have launched a screening app for early childhood learning programs - linking up with our work on the Early Childhood Learning Portal.

Lessons Learned

One of the lessons learned is that civic technologists can provide value not just through arming partners with technology tools, but with knowledge about digital tools.

A large part of our engagement with our partners was simply teaching them that these tools exist and how to leverage them. There are still digital skills gaps in civic organizations and through this project we've been able to advise high capacity organizations like SWOP and LISC-Chicago on technology issues. Between the events we host and the programs we run, we are a great teaching organization.

Lightweight tools that we can deploy repeatedly are an effective way to get impactful results quickly.

Out of all the CivicWorks projects, our use of Textizen showed that we could help make impacts at organizations quickly and without an enormous amount of lift on Smart Chicago's part. Having the 'super' contract with Textizen saved us time on contracting and it helped that we had previous work to show potential partners. It also allowed us to show organizations what was possible with simple lightweight tools. SMS messaging isn't the most fancy of technology tools, but it's something that we've shown can be very impactful.

Product Management is difficult, but do-able

Both projects with FreeGeek Chicago and the Chicago Justice Project contained a lot of product management lift. Product Management, as opposed to project management, is a different skill that required wrestling with a bit of a learning curve. While both products launched and were successful, future civic tech projects should keep in mind the extra time cost. It also helped expose the lack of training in product management skills in the space as a whole.

Products and Projects that perform a function are more impactful than products and projects that tell a story.

The most impactful CivicWorks project will most likely be Smart Chicago's investment in mRelief. mRelief helps to address a problem experienced on the front lines of social service delivery in the city. Our least impactful project was LocalWiki, which while still active, didn't take off to the extent we first hoped.

Sustainability should play a role in future programs.

One of the noticeable effects of the program was that we had many first year partners ask us again for help funding the second year of programming. A few partners indicated that without further funding they would not be able to continue using the service. In contrast, when we partnered with organizations with Textizen we were very up front about this being a one shot deal and it may have helped result in those organizations positioning funds to purchase outside contracts.

mRelief understood instinctively that Smart Chicago couldn't underwrite the program forever and quickly moved to apply for other grants - which they received from a number of sources. mRelief continues to find funding sources to increase their sustainability.

The CivicWorks Project works best as a "force multiplier' for community focused/high capacity organizations working in the trenches to tackle our biggest civic issues.

Smart Chicago has become adept at quickly understanding a technology need and being able to identify potential solutions - and we often have the resources to fund that solution. This has enabled us to provide tools and resources to number of high capacity organizations both inside and outside of the CivicWorks project.

One of our values that Smart Chicago and programs like CivicWorks bring to the Chicago Community Trust and other funders is that we augment the work being done by other civic organizations. When speaking with other civic tech organizations about getting funding, one of the things that's quickly pointed out is the idea of collective impact. Our work in helping grantees already being funded by Knight enhances the power of those grants.

Civic Tech events as learning opportunities, Hack Night and National Day of Civic Hacking

Over the past four years, I've both run and have been involved in a lot of civic technology events. While not all of them have been the same, the most successful events certainly have a common theme: They're learning events.

Civic technologists have to be a little bit of a polymath having to know about two distinct fields. It doesn't help that both of these fields *love* their jargon and work in ways that don't seem obvious to outsiders.

However, the whole polymath thing only goes so far. Civic technology crosses so many different subjects – health, web development, social services, UX design, transportation, DevOps, safety and justice – that people have to spend years learning about it so they can do their jobs effectively. It is incredibly difficult to be an expert on both ends. It requires deep collaboration – build with, not for and the like – to build something that's solves a real problem effectively.

A large part about what makes weekly hack nights so valuable is the chance to learn these skills and the act of putting people in rooms that wouldn't normally interact. What happens when you get a social worker and a UX designer together? Or data scientists and somebody from legal aid? Or a community organizer and an app developer? The civic people get more tech savvy and the tech people become more civic savvy by learning from one another. That's a positive outcome in-and-of-itself, but also empower people to be able to work on a civic technology project when the opportunities arrive.

The way that events are run can also help set the tone for learning. By putting resources online before the event, event organizers can get through a lot of the frequently asked questions. When event organizers call at jargon, it helps set up the expectation that everyone shouldn't assume that people are familiar with all the inside terminology. People are at these events to learn.

This goes double for national events like Code Across and National Day of Civic Hacking. These events attract a lot of first time participants to civic technology events. One of the big lessons I've learned is that how-to guides and simple glossaries are very popular pieces of content. (The glossary that I wrote for this year's National Day of Civic hacking can be found at the end of the book.)

A large part of the strategy around documenting the Chi Hack Night has been to write down the lessons learned so that other people can learn from them. It's not enough to simply know what projects were built or worked on – people need to understand the thought process behind the solution that was built.

As civic technologists continue their work in this field, they must document their efforts in order to raise the overall knowledge of the space. The more civic technologists share, the more likely that we'll avoid repeating work and avoid making the same errors.

Appendix

Appendix A: An Incomplete List of Chicago Civic Technology Groups

As part of our work, the Smart Chicago Collaborative makes a lot of introductions. Chicago has a deep bench of civic innovation talent with multiple organizations working in tandem to use the power of technology to help solve problems. There are also a number of mailing lists and meetup groups that are in the civic innovation space. To help get people who are interested in the space more acquainted, we've put together a list of mailing lists, events, and resources for people who want to get more involved.

Mailing Lists

These mailing lists will help connect you to a broad network of civic innovators.

Code for America Brigade

This is the easiest way to get plugged into the national civic technology community. Once you join this google group, you'll start getting emails from the CfA brigade community and can send out emails to the larger group. This group gets a LOT of traffic, and is a good way to see what's going on.

FreeGeekChicago:

FreeGeek Chicago is a not-for-profit community organization that recycles used computers and parts to provide functional computers, education, internet access and job skills training to those who want them.

OpenGovChicagoland:

This Google Group contains hundreds of Chicagoland residents who have an interest in greater government transparency and civic innovation.

Chi Hack Night:

Subscribe to updates on upcoming Chi Hack Night speakers and events.

Meetups:

These meetings are open to the public and a great way to get involved in Chicago's civic technology community.

OpenGovChicago-land

This Meetup is for citizens who are interested in seeing their federal, state, and local government function more efficiently and responsively. We are inspired by people who are actively building tools and experimenting with solutions along these lines, like the Sunlight Foundation (http://sunlightlabs.com/), and GovTrack (http://GovTrack.us). We believe that open source software practices and internet culture provide good examples of how people can work cooperatively on complex problems to produce meaningful results, but you don't have to be a techie to be part of this meetup. Whether you are an activist or a government employee, or just a hacker looking for interesting projects, come to the Open Government Chicago group to meet other interested people and get involved in making things happen.

Chi Hack Night:

Come join a group of passionate folks working at the intersection of open government, cities, and technology. This is an evening of civic tech hacking, learning, and hanging.

Chicago City Data User Group
You don't have to be a developer to use Chicago's rich sets of data! The Chicago City Data users group is for end users, business users, enthusiasts, students, entrepreneurs, anyone interested in how to use City Data. Our goal to promote civic engagement, innovation, and economic opportunity leveraging The City of Chicago's data.

Connect Chicago Meetup:
People who work to bring public computer centers, community technology centers, and digital literacy programs to the people of Chicago.

If you are part of a facility at http://locations.weconnectchicago.org, this group is for you.

FreeGeekChicago OpenHack
OpenHack is held every Saturday and offers a space to work on software projects with other hackers. Contribute to one of the Supreme Chi-Town Coding Crew projects or use our space to work on your own creation.

Maptime Chicago:
Maptime Chicago is an interactive place to learn about digital mapmaking and GIS in Chicago.

Organizations
These organizations work at the forefront of civic innovation in Chicago.

Chicago Public Library

- Led by Commissioner Brian Bannon
- www.chipublib.org
- Twitter: @ChiPubLIb

One of the first comprehensive technology training programs in the city, the CyberNavigators program at Chicago Public Libraries has led the way in bridging the digital divide in Chicago. The Chicago Public Library also boasts it's own MakerLab complete with a 3D Printer.

City of Chicago Department of Innovation and Technology:

- Led by City of Chicago CIO Brenna Berman
- Data Portal: data.cityofchicago.org
- Twitter: @ChicagoCTO, @ChicagoCDO

The Department of Innovation and Technology (DoIT) was introduced in 2008 to add innovation to the charter of the former Department of Business and Information Services.

As the central information technology organization for the City, DoIT provides a number of technology and telecommunications services to departments, the Mayor, Aldermen, other city agencies, residents, businesses and tourists.

The Department of Innovation and Technology is also responsible for the city's data portal.

Code for America

Code for America is a 501(c)3 non-profit that envisions a government by the people, for the people, that works in the 21st century.

Our programs change how we participate in government by: Connecting citizens and governments to design better services, Encouraging low-risk settings for innovation; and, Supporting a competitive civic tech marketplace.

Data Analysts for Social Good

Data Analysts for Social Good is committed to providing the tools, resources, and community to help social sector professionals use research, data, and analytics think critically about social change.

Data Science for Social Good:

The Eric & Wendy Schmidt Data Science for Social Good fellowship is a University of Chicago summer program for aspiring data scientists to work on data mining, machine learning, big data, and data science projects with social impact.

FreeGeekChicago

FreeGeek Chicago is a not-for-profit community organization that recycles used computers and parts to provide functional computers, education, internet access and job skills training to those who want them.

Knight Lab at Northwestern:

Northwestern University Knight Lab advances news media innovation and education. Developing ideas from experimentation through adoption, the Lab makes technology that aims to help make information meaningful and promotes quality storytelling on the Internet.

Illinois Science and Technology Coalition

The Illinois Science & Technology Coalition (ISTC) harnesses the power of research and innovation to grow the state's economy. We are a member-driven, non-profit entity that drives public-private partnerships between industry, research universities, federal labs and government to increase research and technology-based investment, talent, awareness and job growth in Illinois.

LISC Chicago

Local Initiatives Support Corporation Chicago connects neighborhoods to the resources they need to become stronger and healthier. LISC has a number of programs that intersect in the civic innovation space including the Smart Communities Program and their OpenGov for the Rest of Us Program.

Open City Apps

OpenCity Apps is a volunteer civic apps team that creates apps with open data to improve transparency and understanding of our government.

Smart Chicago Collaborative:

Smart Chicago is a civic organization devoted to improving lives in Chicago through technology. We work on increasing access to the Internet, improving skills for using Internet, and developing meaningful products from data that measurably contribute to the quality of life of residents in our region and beyond.

UI Labs

UI LABS is a Chicago-based research and commercialization collaborative, bringing Universities + Industries together to define problems, design partnerships and deliver scalable solutions to tomorrow's most important challenges.

US Open Data Institute

Moving beyond the "Field of Dreams" model of releasing datasets, we're matching data producers and consumers to create sustainable data ecosystems. We love government. We're collaborating with local, state, and federal agencies (at no cost to them) to help them release data sets and promote their work. Before we even had a website, we had a GitHub repository. We're creating, supporting, and promoting software essential to the open data ecosystem

Civic Startups

Civic innovation in Chicago has reached a point where companies fueled by open data are being formed to provide tools and data that help tackle civic problems. Below are some of the companies that are based in Chicago.

AlderTrak

Insight every weekday on the actions and elections of the 50 aldermen, top city agencies, administrative commissions, their staff and more.

CivicArtworks

Civic ArtWorks makes neighborhood design more accessible than ever. Our online platform enables citizens to share new ideas and concerns about their community with their neighbors. We remove barriers and enable a fluid conversation between community members and our expert planning and design staff. We strive to build trust and encourage greater participation in order to generate a vision for the future that truly reflects the values of the community. This results in diagrams, plans, and illustrations that celebrate the community's input, inspire continued action, and lead to successful implementation.

Chicago Cityscape

Chicago Cityscape is a tool that uses open data to show how, where, and when neighborhood development takes place. Chicago Cityscape was founded by Steven Vance to simplify finding building projects and the companies who do them using the city's copious open neighborhood development data.

CivicWhitaker, Inc

CivicWhitaker, Inc is a civic technology consulting firm based out of Chicago, Illinois specializing in project management, research and writing, digital storytelling event management and community organizing all in the civic technology space. Their biggest client is the Smart Chicago Collaborative which has been incubating the company.

DataMade

DataMade is a civic technology company. We work on projects that make open data, make open source software, and make people more powerful.

Our projects adhere to the Three Os.

- Open data: The project uses data that is entirely open and free to use by anyone.
- Open source: All apps and tools developed are open source and released under an unrestrictive license.
- Open government: The project is civic-minded. It engages and informs citizens about their society, city, or government.

EveryBlock

EveryBlock's goal is to help you be a better neighbor by giving you frequently updated neighborhood information, plus tools to have meaningful conversations with neighbors.

The site is simple to use: create an account and choose which places you'd like to follow — perhaps your home and your workplace. Your custom EveryBlock homepage will be updated throughout the day and show you what's been happening near your followed places, plus what people in those places are talking about. Join in the neighborhood conversation when you've got something to contribute — a question for your neighbors, a news report, an event listing, or just a heads-up about something people in your neighborhood should know about.

Impact Engine

Impact Engine is an investment fund that empowers entrepreneurs, investors, and mentors to make a positive impact on society. Through financial and human capital, we help for-profit businesses address the world's greatest social and environmental challenges.

Kalov Strategies

Kalov Strategies is a consulting firm headed by Josh Kalov. Kalov Strategies is working with the Smart Chicago Collaborative to help open up Cook County data.

mRelief

mRelief simplifies the qualifying process with an easy-to-use form of 7.5 questions through our website application at mrelief.com and our SMS application at 773-377-8946. After the user fills out the necessary information, they are directed to community or government resources based on their eligibility indicators and zip code.

Purple Binder

Purple Binder helps healthcare providers, cities, and non-profit organizations connect people with the resources they need to thrive in their community. We collect community resource data and organize it into directories that stay up to date.

Public Good Software

Public Good Software, Inc was founded on May Day by Chicago startup vets Paul Smith, Dan Ratner, and Jason Kunesh along with fellow Obama for America 2012 alumni Chris Gansen & Aaron Salmon to help people working for the public good with great software. As one of the first Benefit Corporation software startups, we break new ground every day we roll up our shirtsleeves and get down to it.

Rob Paral and Associates

Rob Paral and Associates help organizations deepen relationships with their community.

We do this by providing new demographic, social and economic information, and by determining attitudes, experiences and program usage among the populations served by our clients.

OpenTech Strategies

OpenTech Strategies work to support collaborative open technology ventures through project management, open source licensing, development strategy, developer coaching, release processes, RFP development and more.

SpotHero

SpotHero allows you to search and view available parking options in and around the cities. By reserving your parking online ahead of time through SpotHero, you know exactly how much you're going to pay, and where you're going to park. You can view city parking rates at different city parking garages on the map that are near the address you're searching for.

(SpotHero was also the winner's of Chicago's Apps for MetroChicago Challenge)

Spott

Get more out of your social network by finding the answers you need through the world around you.

- Ask questions and share information with Spott's helpful community using an intuitive location-based search.

- Use your News Feed to browse what's trending in your neighborhood, city, or any other place of interest.
- Privacy is our priority. Spott does not reveal your current location or ask for personal information.

Appendix B: Glossary of Civic Tech Terms

One of the quirks of both working in technology and the civic sectors is that both sectors tend to use a lot of jargon and abbreviations that make perfect sense in context, but can baffle outsiders.

As an organization that believes in collaboration, we end up doing spending a lot of time translating and explaining the jargon.

Our partner, Code for America, has developed a tool for Slack called Glossbot. Glossbot is a simple web app meant to be deployed as a Slack integration. Code for America uses this to define the jargon it comes across as they work so when they're discussing their work everyone can be caught up on the definitions. Of course, this information is also useful for other civic technologists who run across these terms as well.

We've taken a data dump from the bot and edited it to exclude some California things and add some Chicago things. You can see the raw information here. If your team uses Slack, you can also deploy your own Glossbot by forking the code on Github. If you're looking for something in particular, you can also search by hitting (Ctrl+F on Windows or Command+F on a Mac)

- **18f** – A group within the U.S. General Services Administration that builds digital services for government.
- ACCDB – A Microsoft Access database format.
- ADA – American with Disabilities Act; a set regulations regarding hiring practices, building codes, and other regulations aimed at reducing discrimination towards disabled people.

- AFDC – Aid to Families with Dependent Children, a federal assistance program in effect from 1935 to 1996 created by the Social Security Act and administered by the U.S. Department of Health and Human Services that provided financial assistance to children whose families had low or no income. Replaced by the Temporary Assistance for Needy Families (TANF) program.
- AMD – Asynchronous Module Definition; a mechanism for defining JavaScript modules such that the module and its dependencies can be asynchronously loaded.
- ANT -Actor-Network Theory, an approach to social theory and research, originating in the field of science studies, which treats objects as part of social networks.
- API – Application Programming Interface, an interface that exposes the data and functionality of an application to other applications.
- AWS – Amazon Web Services. People can use AWS to host their website on Amazon's cloud servers instead of their own physical ones. Smart Chicago also provides free hosting to Chicago civic apps.
- bikeshedding – Parkinson observed that a committee whose job is to approve plans for a nuclear power plant may spend the majority of its time on relatively unimportant but easy-to-grasp issues, such as what materials to use for the staff bikeshed, while neglecting the design of the power plant itself, which is far more important but also far more difficult to criticize constructively.
- BLUF – Bottom Line Up Front, a paragraph where conclusions and recommendations are placed at the beginning of the text, rather than the end, in order to facilitate rapid comprehension.

- Broadband – Commonly referred to as High Speed Internet; The FCC has recently updated its definition of broadband to 25 megabits per second (Mbps) for downloads and 3 Mbps for uploads.
- CBO – Community-Based Organization or the Congressional Budget Office
- CFPB - Consumer Financial Protection Bureau, an independent agency of the United States government responsible for consumer protection in the financial sector.
- City Analytics Dashboard – A CfA app that displays live activity on a web site via Google Analytics
- Clean – An app built by Code for America's Health Lab to simplify and accelerate the process of applying for CalFresh.
- CMS – Content Management System used by websites which publish a lot of content. A good example of this is WordPress or Centers for Medicare and Medicaid Services
- CRM – Customer Relationship Management used by organizations that have a large sales function or non-profits that deal with a lot of funders or grantees. Salesforce is an example.
- CUTgroup – Civic User Testing Group, a set of regular Chicago residents who get paid to test out civic apps. This program has been duplicated in other cities such as Oakland.
- Digital Services Census – The Local Digital Services Census surveys the quality and usability of the 10 most searched-for city services, as identified from research by the Code for America Tech Team.
- DSL – Domain-Specific Language
- EBT – Electronic Benefit Transfer, the system used in by states for the delivery, redemption, and reconciliation of issued public assistance benefits. EBT cards go by different names in different states.

- EHR – Electronic Health Record
- eRegs - eRegulations, a web-based application that makes regulations easy to find, read and understand, developed by the Consumer Financial Protection Bureau.
- ESL – English as a Second Language
- ETL – Extract, Transform, Load. The process that developers and data scientists use to extract data from one system, transform it, and load it into a different system. Cities like Chicago use this to take data from their business systems and load it into the data portal automatically. For a complete guide on ETL, click here.
- FEC – Federal Election Commission, an independent regulatory agency which regulates campaign finance legislation.
- FPL – Federal Poverty Level
- FOIA – Freedom of Information Act Request; Legal method used to get information from governments. Can be time consuming.
- GDS – Government Digital Service; An agency in the United Kingdom, the Government Digital Service is leading the digital transformation of government, making public services digital by default, and simpler, clearer and faster to use. The GDS provided the inspiration for the US's 18F and USDS.
- GIS – Geographic Information Systems; Software used to display geographic data such as ArcGIS. This term used to reference software made by ESRI, but now can refer to everything from Google Maps, to OpenStreetMap, to CartoDB.
- GitHub – A web service used to host repositories of code and make it easier for developers to collaborate using open source.
- GovDelivery – A marketing and communications firm that works with federal, state and local governments on email marketing campaigns and mailing list management.

- Govtech Fund – A venture capital fund dedicated to government technology startups.
- GSA - the U.S. General Services Administration, an independent agency of the United States government, established to help manage and support the basic functioning of federal agencies.
- HIMSS – Healthcare Information and Management Systems Society
- HIPAA- the Health Insurance Portability and Accountability Act, U.S. legislation that ensures a person's right to buy health insurance after losing a job, establishes standards for electronic medical records, protects the privacy of a patient's health information, and gives patients the legal right to access their own health data.
- IPHI – Institute for Public Health Innovation
- Javascript – Javascript is a programming language used by web browsers. It is not the same as Java which is a whole other language.
- LAF / Legal Aid Foundation – A non-profit organization that often does pro-bono work representing clients who can't afford legal counsel. See LAF Chicago.
- LEP – Limited English Proficient
- MindMixer – A platform allowing local governments to solicit ideas from their communities and support a process for implementing the best ones. A CfA 2012 Accelerator company.
- MOE – Maintenance Of Effort, often refers to requirements that a government must meet to continue participating in a federal program.
- Meaningful Use – Medicare and Medicaid EHR Incentive Programs provide financial incentives for the "meaningful

use" of certified EHR technology. To receive an EHR incentive payment, providers have to show that they are "meaningfully using" their certified EHR technology by meeting certain measurement thresholds that range from recording patient information as structured data to exchanging summary care records.

- NDoCH – National Day of Civic Hacking

NNIP – The National Neighborhood Indicators Partnership (NNIP) is a collaborative effort by the Urban Institute and local partners to further the development and use of neighborhood information systems in local policymaking and community building.

- NSTIC – National Strategy for Trusted Identities in Cyberspace, a White House initiative to work collaboratively with the private sector, advocacy groups, public sector agencies, and other organizations to improve the privacy, security, and convenience of online transactions.
- Ohana API – An API that provides any city or county with an open-source framework for opening up a dataset of community-based organizations, and keeping the information up to date. A 2013 San Mateo Fellows project.
- ONC – Office of the National Coordinator for Health Information Technology. Part of the U.S. Department of Health and Human Services, purpose is to support the adoption of health information technology and the promotion of nationwide health information exchange to improve health care.
- Open Data – Open data refers to data often released by governments that adheres to the 8 Principles of Open Data.
- Open Source – Open source software is software that can be freely used, changed, and shared (in modified or unmodified form) by anyone. Open source software is made by many people, and distributed under licenses that comply with the Open Source Definition.

- OSTP – White House Office of Science and Technology Policy
- PIF – Presidential Innovation Fellows <https://18f.gsa.gov/pif/>
- R – R is a free software environment for statistical computing and graphics. It's used by data scientists to analyze data.
- Ruby – A dynamic, open source programming language with a focus on simplicity and productivity. Often used in conjunction with Ruby on Rails
- Ruby on Rails – Ruby on Rails is an open-source web framework written in Ruby. It was created in Chicago from work that David Heinemeier Hansson was doing for Basecamp.
- SMS – Short Messaging Service, a text messaging service component of phone, Web, or mobile communication systems. It uses standardized communications protocols to allow fixed line or mobile phone devices to exchange short text messages.
- SNAP – the Supplemental Nutrition Assistance Program. This program is sometimes known as food stamps or by the name of the card that people use to get benefits such as LINK or Cal-Fresh.
- SSI – Supplemental Security Income, a Federal income supplement program funded by general tax revenues.
- Streetmix – An interactive street section builder that helps community members mockup the streets they'd like to live on and offer these mockups as future plans for city officials and planners. Built by 2013 fellows.
- Streetsblog – A transportation blog covering transportation and bike issues.
- TANF – Temporary Assistance for Needy Families, a federal program providing cash assistance to indigent families with dependent children through the U.S. Department of Health and

Human Services. Replaced the Aid to Families with Dependent Children (AFDC) program.

- Twilio – A web service used to give apps the ability to send and receive texts.
- USDA – United States Department of Agriculture
- UI – User interface; the parts of the website or app that users use to get the information or results they want.
- UX – User Experience; Generally referring to designing website and applications to make the user experience simple and easy to use. Often means getting apps in front of real users.
- WIC – Women, Infants and Children Program, a federally-funded health and nutrition program that helps families by providing checks for buying healthy supplemental foods from WIC-authorized vendors, nutrition education, and help finding healthcare and other community services.
- WordPress – WordPress is a blogging platform that can be used as a content management system for websites. Its ease of use makes it a popular platform.

The Author

Christopher Whitaker runs the civic technology consulting firm CivicWhitaker, which includes clients such as the Smart Chicago Collaborative and Code for America. He utilizes his experience in government and community organizing to advance civic innovation in Chicago and beyond. Whitaker also serves as the co-host of Chi Hack Night--one of the largest gatherings of civic technologists in the country. He has been honored as a White House Champion of Change for Civic Hacking. He is a graduate of DePaul University (MPA) and Sam Houston State University (BA, Political Science). Previously, Whitaker served with the US Army in Iraq as a mechanized infantryman.

Whitaker tweets regularly at @civicwhitaker and calls Chicagoland home.

The Editor

Daniel X. O'Neil is Executive Director of the Smart Chicago Collaborative, a civic organization devoted to making lives better in Chicago through technology. Prior to Smart Chicago, O'Neil was a co-founder of EveryBlock, where he was responsible for uncovering new data sets through online research and working with local governments. He has worked in the open government/ open data movement since 2004, creating technology, advocating for and writing policy, and working to improve how communities use data to make decisions and improve conditions. Find him at http://danielxoneil.com/.

Made in the USA
Monee, IL
04 November 2026

67384439R00085